Praise for *Five*

"Idiosyncratic . . . emotion-infused . . . deliciously grotesque."
—*The Washington Post*

"*Five Quarts* is a remarkable journey, at once highly erudite and profoundly personal, that leads us through history, religion, science—and our own bodies. Bill Hayes is by turns lyrical, rueful, humorous, questioning, and very moving in this book about himself, about our species, and about our past."
—Perri Klass, M.D.

"Hemophobes beware: there are five quarts of blood in the human body, and Hayes pours all of them into this book. His personal history runs like a river through [it], picking up the flotsam and jetsam of blood lore. . . . His keen perceptions show how the ancient view of blood as the essence of a person's soul still pervades our modern vocabulary and views on the vital fluid. With his strong writing and a unique approach, Hayes satisfyingly addresses this life force."
—*Publishers Weekly* (starred review)

"Sure to make your heart beat faster."
—*Wired*

"*Five Quarts* takes the reader far into the still uncharted country whose center is the human heart."
—*The Buffalo News*

"Bill Hayes's highly original meditation on blood is finally, also, a graceful and subtle love story."
—Richard Rodriguez, author of
Brown: The Last Discovery of America

"Sometimes frightening, sometimes funny . . . full of striking vignettes . . . In its personal riffs on a topic we all know something about, *Five Quarts* will remind many readers of Diane Ackerman's *Natural History of the Senses*."
—*Out Magazine*

"A truly remarkable look inside our own veins."
—*Genre Magazine*

FIVE QUARTS

A Personal and Natural History of Blood

BILL HAYES

RANDOM HOUSE TRADE PAPERBACKS · NEW YORK

2006 Random House Trade Paperback Edition

Published in the United States by Random House Trade Paperbacks,
an imprint of The Random House Publishing Group,
a division of Random House, Inc., New York.

RANDOM HOUSE TRADE PAPERBACKS and colophon are
trademarks of Random House, Inc. READER'S CIRCLE and
colophon are trademarks of Random House, Inc.

Originally published in hardcover in the United States by
Ballantine Books, an imprint of The Random House Publishing Group,
a division of Random House, Inc., in 2005.

Grateful acknowledgment is made to Warner Bros. Publications U.S., Inc.,
for permission to reprint "Blood Makes Noise" by Suzanne Vega, copyright © 1992
WB Music Corp. and Waifersongs Ltd. All rights on behalf of Itself and
Waifersongs Ltd. Administered by WB Music Corp. All rights reserved. Reprinted
by permission of Warner Bros. Publications U.S., Inc., Miami, Florida 33014.

Library of Congress Cataloging-in-Publication Data

Hayes, Bill.
Five quarts : a personal and natural history of blood / Bill Hayes.
p. cm.
Includes bibliographical references and index.
ISBN 0-345-45688-2
1. Blood   2. Medicine—History. I. Title.
QP91.H454 2004
612.1'1—dc22   2004050171

Printed in the United States of America

www.thereaderscircle.com

2 4 6 8 9 7 5 3 1

Text design by Susan Turner

For Steve Byrne

CONTENTS

ONE   GORGON'S BLOOD   *3*

TWO   VITAL SPIRITS   *12*

THREE   BIOHAZARD   *35*

FOUR   BLOOD SISTER   *57*

FIVE   ORIGIN STORY   *84*

SIX   VITAL STAINING   *112*

SEVEN   DETECTABLE   *131*

EIGHT   BLOOD CRIMINAL   *153*

NINE   EXSANGUINATE   *167*

TEN   SHEMOPHILIA   *190*

ELEVEN   BLOOD DRIVE   *212*

TWELVE   BLOOD LUST   *236*

THIRTEEN   MEMORY CELLS   *248*

REFERENCES   *267*

ACKNOWLEDGMENTS   *277*

ILLUSTRATION CREDITS   *279*

INDEX   *281*

Press forth, red drops—
confession drops,
Stain every page

—Walt Whitman,
"Calamus," 1860

Gorgon's Blood

THE FIRST DROP STAINS THE PALE, CLAMMY FLESH. IT'S AS if the skinned potato, not my sliced finger, is bleeding. Were the cut anywhere else on my body I'd have it under the faucet by now or washed with soap, yet I persist in sucking it. The blood is warm, warmer than saliva. This is what 98.6 degrees Fahrenheit feels like on the tongue.

There's always a moment—less than a moment, actually; however long it takes an instinct to fire, as hand flies to mouth—when I think the blood will taste good (an expectation I'd never have, it strikes me, for other bodily fluids): as earthy as cooked beets or sweet as cassis. Wrong again.

Okay, so it doesn't taste good, but it doesn't taste bad, either. If it did, all creatures would be repulsed from licking clean their wounds. Blood's no worse than a lick of sweat yet also not something to be savored. That it tastes faintly like metal, as some people say, is not an undeserved analogy; blood *is* iron-rich. Two-thirds of the body's store of iron can be found there. Others say with great specificity that it tastes like a mouthful of change (*have they tasted mouthfuls of change?*), suggesting, too, that blood is currency, which it certainly is, a donated pint at a blood bank being valued at more than a hundred dollars, according to the FDA. Yet both of these analogies are imprecise, for pennies have a different flavor than quarters, don't they? And there's a world of difference between the lip of an aluminum beer can and a sterling silver teaspoon.

I'm reminded of my high school friend Melaine, now a mother of three, who has worked for twenty years as a technician in a hospital surgical unit and has, to her profound regret, acquired a fine nose for blood. Like body odor, she tells me, everyone's smells different; some blood is pungent and sickening, some almost fruity. She sees blood as a constantly brewing stew, a mix of red blood cells, white blood cells, and platelets, all floating in plasma, the watery medium that carries nutrients to, and waste from, the body's one hundred trillion cells. This basic recipe is then often spiced with medications, alcohol, nicotine, or other ingredients. Melaine thinks that each person's blood is an olfactory signature, which leads me to think that no two samples would taste alike, either.

According to historians, Roman gladiators drank the blood of vanquished foes to acquire their strength and courage (a practice reputedly also followed by the nineteenth-century Indonesian headhunters, the Tolalaki, among other cannibals). But, if true, how exactly did the gladiators drink this blood: in bejeweled loving cups or straight from the jugular? Lapping it off a felled man's chest, perhaps? In any event, why didn't these victorious gladiators drink their *own* blood instead? They were the winners, after all.

Nitpicking about how it was drunk, whose blood was braver, and so forth, however, is missing a larger point. What makes the gladiator behavior truly arresting is this: They didn't fear contact with blood. On the contrary, they gloried in it. Even the spectators were allowed, on occasion, to rush the arena to join blood-drinking free-for-alls. Sated, a gladiator or spectator may have even taken some to later sell. Gladiator's blood, both a cure for certain diseases and a good-luck talisman for new brides, was a valuable commodity. Even so, in those days, the most prized blood of all was not that of a man but of a mythical creature, as found in the tale of Asclepius, god of medicine.

Asclepius, the illegitimate child of a beautiful maiden and the god of light, Apollo, has one of the juiciest backstories in classical mythology. He'd not even been born yet when his mother was slain by his father, who'd become enraged upon learning she had been unfaithful. Apollo snatched the baby from her womb and sent his son to be raised by a centaur named Chiron. From the half-man/half-horse, Asclepius received his medical training, learning to mix elixirs, use incantations, and perform surgeries. From his aunt Athena, the goddess of war and wisdom, he received his most powerful potion: blood from the veins of Medusa, the snake-haired Gorgon whose face turned beholders to stone. A single drop of her blood could either kill or cure a human. If drawn from the left side of the Gorgon's body, the blood brought instant death; if from the right side, it miraculously restored life. This duality was an especially prescient invention of ancient mythmakers, for we now understand in cellular detail how blood can both bear disease with deadly efficiency and save a person's life, as with vaccinations or transfusions. Indeed, Asclepius revived a man named Hippolytus by giving him the precious fluid, which may now be viewed as the earliest—albeit mythical—instance of a blood transfusion.

Asclepius—no fool he—realized there was profit to be made from Gorgon's blood: gold, specifically, which he demanded in exchange for raising the dead. This unethical practice infuriated Zeus, king of the gods, who sent down a thunderbolt, killing the doctor. No matter. Zeus cooled down, realized the overall good Asclepius had brought to humankind, and raised him to godhood. Asclepius went on to father five daughters, all personifications of healing, including Panacea, the divine cure-all. She reigns to this day, in the private domain of my household at least, as the goddess of minor kitchen mishaps. I invoke her name as I wind a Band-Aid around my finger.

In practice, ancient physicians believed that blood was the dwelling place of the "Vital Spirit" that animated human beings—the stream in which emotion, character, and intelligence swam. This life force was thought to be circulated by the heart, which was falsely assumed to function as the body's governing organ (what we now know as the role of the brain). The Roman poet Virgil offered a metaphysical slant. Noting that veins under the skin looked purple, a color lost when blood was shed and a person died, he concluded that the blood must therefore house the human soul—the *purpurea anima*, or "purple soul," as he wrote in *The Aeneid*. It was a reverential perspective almost beyond imagination today, when blood is widely considered hazardous waste material.

I find Virgil's conception marvelous, even though, truth be told, my own veins look aquamarine, not purple. The notion that blood is intricately linked with the soul has a deep resonance for me, above and beyond the fact of my strict Catholic upbringing. Blood, I've found, leaves mnemonic markers at each milestone or phase in life, a way to retrace one's journey. Look back, you'll see. The marks, whether literal or metaphorical, may not be visible at first, like fingerprints before they're dusted, but then, upon second inspection, when the light is shone just so, your whole life looks spectacularly stained in red.

We're born in blood. Our family histories are contained in it, our bodies nourished by it daily. Five quarts run through each of us, on average, along some sixty thousand miles of arteries, veins, and capillaries. Blood permeates religion, as it does the nightly news. Action films are bathed in it. Love songs and poems testify to its thunder. Modern medicines can thin it out, thicken it up, or redirect it to sexually interesting places. With the first shaving nick or menstrual period, blood initiates us into adulthood. It makes us blush, bruise, and go pale. It may leave a trace when a woman's virginity is lost, and a drenching while giving birth. Blood is used to describe a range of emotions: It quickens, races, boils, curdles, runs cold, and sizzles, hot-blooded, under the skin. To say "I feel my blood" is to say one feels alive, vibrant, every cell pulsing.

And yet the mere mention of blood can induce a cringe. Reading about it makes some people squeamish. You can even taste blood without having it on your tongue. It's a taste of, well, wrongness, I suppose—an emotional "taste" that things aren't as they should be. At such moments one turns into a kind of temporary synesthete, a person whose senses are commingled. Synesthetes taste sights, see tastes, feel colors, hear shapes. An E-flat triggers a visual field of triangles, say, or pain manifests as a blue aura. For the Russian novelist Vladimir Nabokov, as idiosyncratic in his synesthesia as in his writing, every letter of the alphabet radiated a precise color; for example, *O* was the hue of an "ivory-backed hand-mirror." His wife, Vera, who also had synesthesia, would likely have disagreed, for two synesthetes' perceptions are almost never the same. An ivory *O* to one is black shoe polish to another.

This condition is rare—as few as ten people in a million have synesthesia, according to a recent American study, a figure that makes the pairing up of Vladimir and Vera even more extraordinary. Yet many of us, I believe, experience a similar phenomenon upon seeing blood. Blood is warmth to some; an overheated room. To others, blood is noise; a rabbit's racing heartbeat. But I don't see it that way. To me, blood is silence. It is slow motion, the pause between seconds. It is a sharpness, dry air, clarity.

Blood also marks the divide between me and my partner of fourteen years, Steve. He has HIV, a fact I've known since our first date. Steve's always been extremely cautious about blood—overly so, I've thought at times. If he nicked himself shaving, for instance, he wouldn't allow me within kissing range of his face. At those moments, I tended to act blithely unconcerned. Against his wishes, I might plant a kiss on his forehead—"See, no harm done." I've never wanted Steve to think I was afraid of him.

I've seen him with a bloody nose before, but not with a serious, bleeding wound until a short time ago. While he was stocking food on the bottom shelf in our pantry, the iron toppled from the top shelf, a fall of about three feet, hitting Steve, tip-first, on the head. I heard his *oophf!* of pain and surprise and watched him stumble out of the pantry, trailing blood like beads of a broken necklace. My first impulse, *Help him,* was instantly followed by a second, seemingly contradictory one: *Protect yourself.* With blood dribbling down Steve's face, mine went pale, I could tell, and I froze.

I stood there at least long enough for our eyes to meet. I

looked stunned, no doubt, if not plain guilty; I'd been the last to use the iron and had left it too close to the edge. But concern for Steve broke through my self-absorption. I pivoted and grabbed paper towels as he said, calmly yet forcefully, "Put . . . on . . . gloves." His extended hand established an invisible force field between us, a danger zone I could not enter. Though I didn't have any obvious open cuts through which I could get infected, I knew it was better to be prudent. I dropped the towels and scrambled to the bathroom closet, returning with a pair of disposable latex gloves, gauze, and disinfectant.

Steve sank to the kitchen floor and sat there, eyes closed, gingerly assessing the damage with his fingertips, a slow spider's crawl to feel if his skull was cracked. His hand came away sopped, but, he told me, there was no fracture, just a gash. Gloved, I blotted it with gauze and examined the scalp beneath his sticky flattop; the cut didn't look deep enough to require stitches. It stopped actively bleeding in minutes, in fact. Steve was even able to joke that his Mega Hold hair gel must have blunted the impact.

Though this isn't a pleasant admission, after Steve was cleaned up, I realized I'd found it thrilling to see his blood in such quantity—on his brow, staining paper towels, in a spattering on the floor. Its luminous red color was like molten lava, both terrible and beautiful. Until that day, it had been an abstraction to me, something drawn and analyzed behind a closed office door, mere numbers on a lab report, or a bluish fluid glimpsed through his veins. I was used to seeing Steve's semen when we had sex; that didn't scare me. But his HIV-positive blood was altogether different, seeming far more dangerous by being rarely seen. How well I knew its capacity to bring good or bad upon us—twin shades of the Gorgon. It made him sick. It helped him fight back. It kept him healthy. It could kill.

Returning to the myth I'd known so well as a kid, I of course remembered the basics: To free his captive mother, the hero Perseus had to deliver the severed head of the Gorgon Medusa, who was as deadly to look at as she was ugly. I was struck, though, by the myriad details I'd forgotten. The goddess Athena joined Perseus as his battle adviser, for instance, and others in the pantheon offered their aid. Winged sandals and a helmet of invisibility were gifts of the Nymphs. A sword made of adamantine, a material as hard as diamond, was bestowed by Hermes.

Perseus with the severed head of Medusa

With Athena by his side, Perseus flew to the Gorgon's domain, landing amid a gruesome rock garden of men the monster had turned to stone. Each statue, a failed slayer forever frozen in horror, spoke to the utter hopelessness of Perseus's task. A cold chill passed through him; for a moment he, too, was frozen, but he pressed on. Carefully, quietly, he made his way into Medusa's lair. Fortune smiled upon him—the monster was sleeping! To protect Perseus should she awaken, Athena covered Medusa's face with her shield. And with a single stroke of his blade, off came the snake-haired head.

A crimson puddle formed at Perseus's feet. As Athena set to work collecting the magical blood she'd later give to her nephew, Asclepius, the pool spread. Laying down his sword and helmet, Perseus reached for the severed head, being careful not to look at it, and stuffed it into a magic satchel. In the blood below him, Perseus saw his own reflection, the face of a man who'd done the impossible and survived.

It is this conquering of one's worst fears that I've always found satisfying, in tales spun on a grand scale and, more so, in those played out on the mortal plane. I've certainly known times when I've been paralyzed by fear or self-doubt but was still able to dig down and push through. These kinds of struggles almost always have unexpected results, as Perseus also discovered. Before his weary eyes, the thin layer of Gorgon's blood began to ripple and, suddenly, an immaculate, white-and-gold creature leapt out, the winged horse Pegasus. It spread its magnificent wings and flew off to join the Muses, where it would come to serve as the spirit of poetic inspiration. But the pool of blood held one last surprise. Out crawled a foul giant brandishing a sword, the warrior Chrysaor, who'd go on to sire a three-headed monster and a man-eating daughter. The story of Perseus and the Gorgon has thus inspired mine, a personal odyssey through the history of hematology, from the classical age to the modern and, along the way, into my own past. With blood as the mirror, I look at my life and see what emerges.

TWO

Vital Spirits

AT THE GYM, TWENTY-FOUR MINUTES INTO A RUN, THE
speed set at 7.1, incline at 5 percent, my feet barely
tap the treadmill as I hit mile three. I am in the silence between songs on my Walkman, then Björk's
"Enjoy" begins. Shoulders thrown back, heart pound-
ing, a shiver runs up my spine as I push against the
resistance. All five quarts of blood pulse and churn
and, without being able to pinpoint the exact moment, I transcend, riding a wave of endorphins.
Outer and inner rhythms merge and I turn up the
volume. I enter a kind of cardiovascular nirvana

where my soul seems to burst from my chest. Music and breath, blood and sweat, this is the closest I come to God.

Even after stepping off the machine, I am six inches above ground, as clear and weightless as the air I'm devouring. This divine conjunction of sound and movement comes only once or twice a month, if I'm lucky. Far more than an "endorphin high"—that euphoric sensation exercise can induce—it is difficult to re-create. And when I get into this groove, I cannot say for sure where it comes from, whether I'm drawing upon energy from deep within my being or responding to a force outside myself.

Steve knows the exact sensation, though he has a harder time nowadays physically attaining it. A man who's nothing if not fearless with his metaphors, Steve told me a couple of years back that one of the clearest descriptions of this feeling, this quasi-spiritual energy surge, could be found in a comic book. I was dubious, but as he told me about a character called the Flash, I began to see how perfectly it did fit. You see, he explained, this superhero is able to run superfast because he taps into a field of energy called the Speed Force.

I wanted to know what it looks like in the comic, how the Flash finds it, that sort of thing.

"It just exists," Steve said, matter-of-fact. "He's able to 'feel' it." He smiled and shrugged, adding, "He can also share his speed with other people, share his power."

The strength Steve and I have drawn from each other over the past decade has certainly gotten us through some rough times as he has battled AIDS. Of course, love and support are just a part of our arsenal. His survival has depended, to a great extent, on superb doctors. And the "miracles" we've experienced in our life—the powerful new drugs approved just in time, the wholly unexpected positive blood work, Steve's Lazarus-like return from dire illness—seem more a product of the pharmaceutical industry than the intervention of a divine being. While things are currently not great, they're not terrible, either. We're in a holding pattern of hope, anxious but not desperate. We have placed our faith in science.

But faith does not preclude questioning. I've never missed joining Steve at a doctor's appointment, and we always arrive with a list of questions to raise, whether about symptoms, drug side effects, or medications in the pipeline. The three of us go through Steve's blood work test by test to assess how his regimen is working. When his doctor recommends a new drug, we then do our own research in treatment journals and online before filling the prescription. The scrutiny to which new medications are subjected today is easy to take for granted. It's mind blowing, by contrast, to look into the history of a long-lived remedy for which the most extravagant claims were made when no conclusive proof of its value ever existed: bloodletting. The practice of withdrawing blood to treat a spectrum of ailments, everything from insomnia to hemorrhage, only died out in the United States in the 1920s. It had endured throughout much of the world for more than twenty-five centuries—twenty-five centuries!—making it, in my view, the longest-running clinical trial in medicine, one that involved millions of patients and persisted on nothing but anecdotal evidence.

The earliest and most influential surviving texts on bloodletting were written by a Greek doctor named Galen (A.D. 129–200), who began his career tending to wounded gladiators and rose to become the Western world's supreme authority on medicine. Galen's views were considered medical gospel for fourteen hundred years, and I can understand why. He makes an illogical practice sound downright reasonable. His writing voice is so clear and commanding, it is almost a summons.

GALEN FIRST MADE A NAME FOR HIMSELF AT THE COLISEUM IN Pergamum, a small kingdom in what's now western Turkey. There, as the chief physician to the gladiators, Galen was the ancient equivalent of an ER doctor, treating the freshly butchered in his trauma ward in the basement of the stadium. Just as a modern physician might hear the distant wail of an ambulance and know a body was on its way, Galen may have had his own alarm system in the collective gasps and muffled yelps of the crowds overhead. This gifted twenty-eight-year-old, who'd begun studying the healing arts at fourteen, had an impressive first year on the job: Not a single gladiator died of injuries, which was unheard of, given that the combatants fought with brutal-looking tridents and two-foot-long swords. The position also gave him an unprecedented scientific advantage: rare, close-up views inside still-living bodies.

Neither of Galen's most distinguished predecessors, Aristotle (384–322 B.C.) and the Greek physician Hippocrates (460–375 B.C.), had ever dissected a human body. And those who had performed dissections noted that the corpse's arteries were empty (because, as a modern pathologist would explain, blood drains back into the veins once the heart stops). This led the great thinkers to the false conclusion that the arteries contained only an air-like substance. Hence *artery*, derived from a Greek word meaning "air duct." Galen, with virtual human vivisections at his disposal, was able to correct this error. Clearly, the arteries were filled with blood. Further, he accurately traced the course of blood from the right side of the heart to the left via the pulmonary artery and lungs. The most crucial discovery, however—that the arteries, veins, and heart work together as a circulatory system—would not be made until the 1600s.

Despite his new insights, the foundation of Galen's medical philosophy did not change. He embraced the ancient Hippocratic theory that a person's state of health was determined by four fluids, or humors—blood, phlegm, yellow bile, and black bile—which had to be in perfect balance within the body for good health to prevail. When one or more of these fluids was overabundant or insufficient, the body's inner scales tipped and disease resulted. The humor called black bile deserves special mention because, as modern medical historians point out, it is a fictitious substance, though Galen firmly believed it existed. Conceptually, there *had* to be four humors—not three, not five—because the body was seen as a microcosm of the universe, which was organized in patterns of four: the basic elements (earth, air, fire, water) and their cosmological correlates (the Earth, sky, sun, sea); the four seasons; and what were known as the Aristotelian "qualities" (cold, damp, hot, and dry). All of these aspects were interrelated, sometimes overlapping, in an elaborate system that Galen would later codify in his medical books. Phlegm, for example, was cold, damp, watery, and wintry, while blood was hot, damp, air-, and spring-like. That the microcosm mirrored the macrocosm was a defining part of Galen's diagnoses. After patching up a gladiator who'd lost a large amount of blood, for instance, he would likely have prescribed a huge increase in the man's dietary intake. In the doctor's view, the liver—the body's hot, damp organ—was "the principal instrument of sanguification"; it converted digested food into blood. The patient, by gorging himself, would restore his humoral balance. Were another gladiator unable to fight because of, say, a high fever, Galen would have diagnosed "plethora," an excess of humors in his body. As treatment, Galen would have turned to copious bloodletting.

With only a slight shift in thinking, I can understand Galen's rationale. Since physical appearance was one of the few diagnostic tools available to him, it's not wholly absurd that Galen would've concluded from a flushed, fevered face that the patient was suffering from superfluous blood. Or, similarly, that a sallow complexion meant that the gallbladder—the hot, dry organ—was producing too much yellow bile. To have explained to Galen that things floating in the blood actually caused disease would've made as much sense to him in his day as saying his fax needed paper. In second-century medicine it wasn't the quality of blood that was at issue, but the quantity. One could have too much of a good thing.

Doctors in Galen's time had other methods for cutting off the body's production of fresh blood: getting the patient to stop eating for several days or inducing vomiting (after all, who doesn't look a bit bloodless after throwing up?). But Galen preferred the immediacy of *venesection,* opening a vein. He used a pointed, double-edged metal scalpel, an instrument of the period called a *phlebotom* (from the Greek *phlebos,* for "vein," and *tome,* meaning "to cut"), from which came the medical term for drawing blood, *phlebotomy.* Few of these original devices survive to the present day, but one found in the ruins of Pompeii was probably typical: a slim, three-inch-long bronze blade mounted in a decorative handle.

When Galen put a phlebotom to a vein, he was continuing a medical tradition that had existed since at least 2500 B.C., the approximate date of an Egyptian tomb painting showing a noble being bled at the foot and neck, the earliest-known depiction of phlebotomy. Of course, this is not meant to suggest that there hadn't been two and a half millennia of naysayers, just that scant records survive. What is known is that, from Hippocrates's day forward, the critics of bloodletting were as passionate as the advocates. Galen discovered this firsthand when he moved to Rome, the big city, in the year 162. A local celebrity in Pergamum, the thirty-three-year-old found that his specialty made him an outsider here, and he immediately butted heads with the Roman medical establishment. His biggest opponent, strangely enough, was a man five hundred years dead, the Greek physician Erasistratus (300–260 B.C.), who'd taught at the celebrated medical center in Alexandria and had been vehement in condemning venesection. He had a legion of vocal followers, known as Erasistrateans.

In a colonnaded hall opposite the Forum, Galen joined others in giving public lectures during his first year in Rome. In the spirit of self-promotion, he held forth on a range of topics, including, notoriously, his enthusiasm for bloodletting and his disdain for the teachings of Erasistratus. The Erasistrateans in attendance were not pleased, although their heckling probably just drew larger crowds. Galen relished stirring up controversy. He was also eager to drum up business. Unlike the majority of Roman physicians, he proclaimed himself to be above all sects, the leader of his own school of thought. He ingratiated himself to the audience, which swelled over time as word spread of the brash, charismatic healer. Bored with chariot races at the nearby Circus Maximus, dignitaries dropped by to see the theatrical young man who, for example, demonstrated vivisection with a squealing pig. Shorthand writers transcribed his speeches and, within months of his arrival, Galen's views on bloodletting were published in the book *Against Erasistratus,* the title alone like a gauntlet thrown down.

A modern reader of the book and its sequel, *Against the Erasistrateans Dwelling in Rome,* gets a vivid sense of this renegade doctor, a man whom one nineteenth-century historian described as a "quarrelsome, self-willed spiteful brawler, who goes for his adversary foaming at the mouth." Nonetheless, Galen's words lift from the page and, like a gentlemanly participant in a cross-time debate, he reiterates his opponent's positions before dissecting them, even quoting the dead Erasistratus verbatim. This is fortunate since Erasistratus's words survive only within the works of others.

Galen starts off on a conciliatory note: The two physicians would not have disagreed, he admits, about what causes illness—humoral imbalance—or about the final goal of treatment, the "evacuation" of the plethos. They parted ways when it came to the means. "The easiest and promptest course of action is to open a vein," Galen states. "In this way, we evacuate the actual inflammatory materials themselves. And nothing else." It's a remedy that was "esteemed by the ancients." By contrast, the main Erasistratean solution, starvation, "apart from the long time it requires, evacuates the whole system indiscriminately." Even now, I can almost hear Galen's *tsk-tsk,* as well as the grumbling of the Erasistrateans gathering at the back of the crowd.

In addition, starvation, Galen warns, is accompanied by a host of evils: severe fatigue, nausea, heartburn, constipation, and perhaps the most serious side effect, turning the other humors "bilious and painful," thereby exacerbating rather than alleviating the imbalance. Pausing, I imagine, to let the crowd absorb these grim facts and to allow the stenographers to catch up, Galen then adds, "Yet Erasistratus sees none of these." He and his adherents are "like blind men, who although a smooth, broad, and direct road is near, often take a narrow, rough, and long one, and go by a circuitous route."

A surgeon by training and temperament, Galen next attacks their use of powerful laxatives and purgatives as a foolhardy reliance on fate. "The flow to the stomach cannot be stopped in the way you can immediately put an end to the bleeding by putting your finger to the divided vein." He then brings this argument home: A "grave disturbance of the entire body [may occur] as a result of being evacuated either insufficiently or to excess," whether it be loss of consciousness or "pulselessness." Indeed—pause for dramatic effect—"the ultimate misfortune often ensues in this state."

What's fascinating is that Galen's own last remarks could equally have been used to denounce his beloved bloodletting. Erasistratus held that it was impossible to determine precisely how much excess blood a patient had and therefore how much should be let. Further, he had seen the handiwork of incompetents who'd slashed through tendons, nerves, and arteries in their hunt for a vein, leaving behind lifelong damage if not death. The practice was so fraught with risk that Erasistratus viewed bloodletting as tantamount to committing murder. But Galen felt himself beyond reproach. He well understood that a patient could die quickly if an artery were severed, which was the primary reason he rarely attempted opening these thicker vessels with their greater blood volume. (They were also harder to access, anatomically speaking.) As for patient safety, Galen again placed himself above Erasistratus, who "paid little attention to examining patients, but stayed at home and wrote down mere opinions." Adamant that only an experienced physician should perform phlebotomy—he, of course, being best suited—Galen advised cutting parallel to a vein, never across it, and keeping puncture holes small. Bloodletting sounds ghastly, but, in fact, the amount Galen drained at one sitting was modest—about a pint at most, no more than you'd give today at a blood bank. However, the doctor often chose to repeat the procedure day after day after day.

Galen saw no stronger argument for venesection than in the seemingly spontaneous bleedings orchestrated by "Nature." He cites nosebleeds and menstruation as ways the body restores its humoral balance: "Does [nature] not evacuate all women every month by pouring forth the superfluity of the blood?" As a matter of fact, modern medical historians speculate that menstrual bleeding probably not only provided the initial inspiration for drawing blood but also helped reinforce its supposed benefits. After all, as my five sisters and numerous female friends attest, they feel great after their period is over.

"But enough of women," Galen says dismissively to the crowd. To further his argument, he calls on a most unlikely ally: the hemorrhoid. "Come now to consider the men, and learn how those who eliminate the excess through a hemorrhoid all pass through their lives unaffected by diseases." That bleeding hemorrhoids are to be appreciated and even encouraged, I daresay, surprises most of the assembled sufferers.

His provocative comments did not go unanswered. The Erasistrateans shot back rebuttals and insults, and, outside the hall, rivals spread malicious gossip. In time, Galen began to fear he'd be poisoned by his enemies. For his own safety, after barely a year in Rome, he stopped publicly antagonizing opponents. Galen turned to writing, producing books in which he both challenged the conclusions of the rival sects and put forth his own findings. He began to tutor privately, and his medical practice flourished. Among his patients were members of the royal court, including a son-in-law of Emperor Marcus Aurelius. In less than four years after his arrival in the city, Galen was consulting with the emperor himself. When Marcus Aurelius's small son fell ill, Galen cured him, earning not only the father's complete trust but also the title of Emperor's Personal Physician. Chief among his duties: concocting antidotes for poisons that might be used to assassinate the ruler. Galen made himself an expert at creating a wide range of treatments from herbal extracts. These came to be called *galenicals,* a term still used for medications made purely from botanical ingredients.

The same son whose life Galen had saved, Commodus, abruptly fired the doctor upon succeeding his father as emperor at age nineteen. Commodus (portrayed by Joaquin Phoenix in the 2000 film *Gladiator*) was a brutal, decadent tyrant responsible, historians agree, for leading the Roman Empire into steady decline. Galen remained in Rome but kept a low profile, quietly continuing his scientific work and his ceaseless writing. He wrote eighteen books on the subject of the pulse alone, as well as tomes on fever, anatomy, the nervous system, nutrition, and philosophy. Most of his original manuscripts were stored for safekeeping at a temple guarded by priests. Unfortunately, in the same year as Commodus's assassination, A.D. 192, a fire destroyed the temple and half of Galen's life's work. About 120 books survived.

With Commodus gone, the doctor, now in his midsixties, slowly reemerged and again began making house calls to the royal palace. Instead of resuming his former duties, however, Galen realized that the new emperor and his wife had other ideas. The queen, aware of his genius for creating botanical remedies, commanded that he concoct equally miraculous beauty potions. Though the assignment was beneath him, Galen must've felt he had no choice. He whipped up extractions to turn black hair golden, face paints for eye shadow and rouge, perfumes and pomades. He was the Max Factor of his day, minus the screen credits or royalties. But he made the best of it, funneling what he learned about cosmetics into further research in pharmacology and, of course, into new books.

BY THE TIME OF HIS DEATH AT AGE SEVENTY, NO ASPECT OF THE human body had escaped Galen's scrutiny. No organ remained unidentified. No ailment could not be cured by his means. He left behind detailed opinions and instructions on everything from venesection in children to mixing up the perfect eyeliner. He did such a thorough job that, in the fourteen centuries following his passing, few dared challenge Galenism, as his teachings came to be known. His reach even extended to the Eastern world, where his books were translated into Arabic in the ninth century. Whereas Asclepius was the mythical god of medicine, Galen was close to the real thing. Indeed, in the early Middle Ages, church leaders declared his work to have been divinely inspired and thus infallible, dubbing him Galen the Divine. To oppose him was blasphemous, punishable with death—burning at the stake. All of which now seems ironic because Galen had never been a religious man and had, in fact, championed the value of scientific experimentation.

Still, the question remains, why did Galen hold sway for so long? Why did he, and not others, endure? Part of it was luck, to be sure, plus sheer volume. While half of Galen's original works were lost, his surviving literary output—some 2.5 million words, according to one count—overwhelmed his competition. But most of all, Galen's influence rested on his blazing self-confidence. As medical historians have observed, Hippocrates, by contrast, acknowledged the potential for error in his work, areas he did not yet fully understand. But not a whit of doubt appeared under Galen's name.

At times I find his arrogance galling (a condition that Galen's fifteenth-century followers would've diagnosed as an enflamed and leaky gallbladder, the repository of this particular emotion). But as a writer, I can't help being impressed by the faith he inspired in others through the force of language. It's all the more amazing to me because Galen was so often dead wrong.

Truth be told, I find scientific blunders as fascinating as the great discoveries. It's the main reason I enjoy reading archaic medical texts, a trusty form of time travel I undertake at libraries. Being in the position of knowing more than Galen did is satisfying, I will not deny. It's also sobering. Two decades from now I'm sure I'll look back and shake my head, amazed at the things Steve and I once did in the name of cutting-edge science. But there's something else. Looking back at Galen looking forward, I am touched by his efforts to treat deadly illnesses, to alleviate suffering, however futile. Finally, he is most impressive not in having come up with so many answers but in taking on so many big questions, such as, What is the essence of life? What makes us human? Galen believed the ingredients were in the bloodstream, where a trio of incorporeal "spirits" flowed. (By contrast to the groupings of four used at that time to describe the inner and outer workings of the universe—the humors, the qualities, the elements—the spirits came in threes, reflecting the tripartite division of the soul theo-

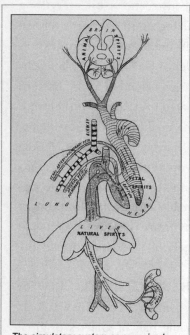

The circulatory system as conceived by Galen

rized by Plato.) The first two ran in the dark, purplish blood of the veins: Natural Spirits, brewed in the liver, providing the body's mass; and Animal Spirits, fired in the brain, producing movement. Completing the trinity were Vital Spirits, the essence that separated human beings from animals. In its fleet passage through the heart, the scarlet arterial blood was imbued with this zest, which disseminated warmth and verve throughout the body. In Galen's reckoning, the spirits did not intermingle; the veins and arteries were separate streams.

As scientific bunglings go, Galen was in good company. No less a genius than Leonardo da Vinci (1452–1519) made some spectacular ones in his notebooks of anatomical drawings. Leonardo, who dissected cadavers and sketched directly from them, set out with the express goal of being faithful only to the evidence of his eyes. Unlike Galen, he held a lifelong aversion to verbiage and believed that drawing was the only uncontestable means of expression. That being said, Leonardo still could not escape Galen's lingering influence. This groundbreaking artist who rendered with astonishing accuracy the chambers of the heart, for example, and the fetus in utero, nevertheless added fictitious plumbing to the human body—canals, ducts, and veins—to accommodate humoral theory. Likewise, he drew the spleen cartoonishly large, proportional to its inflated role in secreting the illusory black bile. Another fallacy perpetuated by Galen and then by Leonardo was the kiveris vein, which resolved the biological puzzle of why pregnant women stopped menstruating. The answer: Menstrual blood was converted into mother's milk, of course, and this "milk vein" conveyed it from the uterus to the breasts. Uniquely male anatomy was fictionalized, too. In cross sections of the penis, Leonardo added a phantom vein for "vital impulse," the life-giving *oomph* ejected alongside sperm. Of all Leonardo's fabrications, the cleverest, I think, was his explanation for crying. A slender vessel carried tears from the heart, the organ of the emotions, up to the eyes. (One last phantom vessel of note is the *vena amoris,* the "love vein," first described by the ancient Egyptians and absorbed into Christian ceremony in the fourth century. The *vena amoris,* it was believed, carried blood straight from the fourth finger of the left hand to the heart, which accounts for the enduring custom of wearing one's wedding band on this finger.)

I take it that, in the past, it was easier to believe in the unseen, the unproven. To feel certain that universal forces were reflected in the human body. Modern medical technologies all but dash these notions. Still, I share Galen's and Leonardo's conviction that real answers can be found within, even though they don't show up on MRIs, CAT scans, or blood assays. I do pin my hopes on intangibles, from my simplest expectations to my most fervent dreams.

But that faith is tested. Every three months Steve gets his blood drawn. Where, historically, the removal of blood was a remedy for disease, in modern-day phlebotomy it's done for diagnostic purposes and to gauge how a treatment is working. Nothing brings Steve and me more down to earth than the reality of bad blood counts. Nothing launches us higher than when the results say his virus is "undetectable," which has lately been the trend. That it's *found* to be *undetectable* is a delicious oxymoron. What this means technically is that so few "copies" of HIV exist in his bloodstream that it cannot be measured. The virus, in essence, is neutralized, not rapidly replicating and therefore less capable of inflicting harm. It also speaks to the limits of present technology. The amount of virus Steve has simply falls below the radar. While, in truth, an undetectable virus is as much of an illusion as Galen's Vital Spirits, the word nonetheless carries tremendous weight. Right now, it's the closest we have to "cure."

To the extent that one can manage a life-threatening disease, Steve has been unusually successful, adhering without fail for years to a difficult regimen of pill taking. "Comply or die" is his motto, though I doubt ACT UP will make a T-shirt of it. Since starting on protease inhibitors in late 1995, he's had no AIDS-related illnesses, although painful nerve damage caused by earlier drugs (a condition called peripheral neuropathy) persists. Along with his meds, he does everything he can to keep his mind, body, and blood as healthy as possible. I would fault his one bad habit—an overly fond attachment to Diet Mountain Dew—were I not similarly addicted.

Steve has blood drawn about two weeks before his scheduled doctor's appointments, so that the results will be waiting when he arrives. The logistics are no more complicated than that. It's a cakewalk compared with how convoluted taking blood became in the Middle Ages. In the fifteenth century, for example, the process depended on a fussy convergence of factors, owing more to celestial bodies than to the patients'. A physician took into account the influence of the sun and moon, the principle being that earthly tides were reflected in the flow of humors. Signs of the zodiac were, in turn, linked to body parts. Aries, for example, was matched to the head, so in late March blood would only be let from the temple. In time, the calculations got so Byzantine that a doctor had to rely on bloodletting calendars and handheld devices adopted from astronomy to determine the right moment to snip a vein.

Steve's quarterly blood draw has always been a joint ritual, in which I drive the car and provide companionship. After all these years, it's still nerve-wracking, yet, thankfully, we've come a long way from the time when the results were so consistently poor that his doctor stopped testing his blood altogether. For the past ten years Steve has used a lab called Immunodiagnostic Laboratories, located in a downtown medical building. The door to IDL looks like a private eye's office in a film noir: hand-painted black lettering on thick mottled glass, a dark oak frame, a well-worn mail slot. Unless you had business within, you'd be hard-pressed to guess what's behind it. The tiny waiting room inside is narrow and dim, an overheated den dominated by old magazines.

SAME MAGAZINES, DIFFERENT LOCATION: THE OLD-FASHIONED BAR-bershop, alive and well today in small-town America, with its trademark candy-cane pole out front. This, like IDL, is a descendant of medieval bloodletting establishments. Back in thirteenth-century Europe, venesection was the specialty of the "Surgeons of the Short Robe," who were also called barbers. ("Surgeons of the Long Robe" performed more elaborate operations.) Barbers also cut hair, stitched minor wounds, gave enemas, and extracted teeth. At this stage of worldly enlightenment, it was considered healthy for an individual to be bled a couple times a year just to remove the buildup of toxic humors. Think of it as opening the window of a stuffy room. To advertise his services, the barber posted a striped pole outside his door. When I was a boy, this pole was a reminder of the candy I'd earn if I didn't fidget. The truth is far more ghoulish. The red stripe symbolized blood; the white stripe, the bandages; the blue stripe, the vein; and the pole itself represented the stick the patient gripped to facilitate blood flow. Barbers continued to perform venesection up through the seventeenth century, and early colonists transported the practice to America.

Although Steve never makes an appointment for a blood draw, he almost never has a wait at IDL. He needs to fast for certain tests, so we're there first thing in the morning. Seriously groggy, Steve is like a big sedated dog that's followed me into the waiting area. The receptionist's window slides back, a courteous hello rings out, and Steve hands over his lab write-up. Usually, once he's called inside, I sit down and use the time to catch up on ancient celebrity gossip. Today, with the permission of lab manager Rosemary Cozzo, I'll be a fly on the sterile office wall.

IDL's inner offices are as bright as a new refrigerator and divided into cool, white compartments. Steve, a foot taller than Rosemary, squeezes into one of the draw-station chairs as she studies the lab form. There's not much room for spectating. I could easily get in her way. Fortunately she is someone over whose shoulder I can actually look, my five-eight to her five-one. As if someone has just said *Go,* she starts plucking empty vials, three purples, two yellows, and assembles the other equipment she'll need.

Rosemary, a licensed medical technologist in her late fifties, could illustrate the dictionary definition for *nurse* (see also, *efficiency*). In her starched white, monogrammed lab coat, skirt, and low pumps, the only thing she's missing is an old-fashioned nurse's cap bobby-pinned to her no-fuss hair. She has a heart-shaped face and a warm smile. As she snaps on latex gloves, my eyes are drawn to a prominent vein on her left temple, a blue squiggle under her ivory skin. If Galen were here, I can't help thinking, he would want to bleed from it. He devoted tremendous attention to mapping the body's veins as sites for letting, everywhere from behind the ears to the roof of the mouth to the ankle. These days, by comparison, blood is almost always drawn from a vein in the crook of an elbow. If one is difficult to access—say, if a patient is obese—a vein on the hand or foot might be used. There's no such problem with Steve, who has lean, muscular arms and the big, ropy veins of a gladiator. Rosemary looks pleased, as does Steve, though for a different reason. Some days a newly trained staff person draws his blood, and that rarely goes well. Even before the needle is unsealed from its packet, Steve's told me in the past, he can tell just how new a novice is.

"A newbie looks at your arms with a great deal of indecision, as if weighing a dozen options. But there are really only two: right arm or left," he's pointed out. "And when the person starts poking at your veins, self-narrating about which one looks best, this one, no, maybe this one, that's when I think, *This isn't gonna be pleasant.* It's also a bad sign when the person rubs the alcohol on your arm like they're trying to remove a tattoo."

There have been mornings when I've been able to tell how rough a blood draw was by how damp Steve's T-shirt is. "Three times," he'd say, for instance, joining me in the waiting room. "Three times to get the needle in right." Or sometimes he'd say nothing and just flash me his two bandaged arms.

With Rosemary, no uncertainty is betrayed, and this translates into a confident spearing of his vein. "It's like cracking an egg without smashing the shell or breaking the yolk," Steve has said; "swift and decisive." I now watch her technique. The needle and housing come packaged like a vending-machine sandwich; Rosemary pops open the seal. She then ties a tourniquet to his left arm, swabs the distended vein, and, in the blink of an eye, slips in the three-quarter-inch needle. Steve doesn't flinch. (I do.) He's had this procedure done at least fifty times in the past dozen years; he's used to steeling himself against discomfort and potential bad news. That he's had to learn this skill, I find heartbreaking. In this context I appreciate Rosemary's gentleness and competency. Unlike some phlebotomists, she always uses a "butterfly," a needle stabilized against the skin with tapered "wings" and connected to a narrow, eight-inch tube. At the end of this tubing is a barrel into which consecutive vials are inserted. The vacuum in each vial sucks Steve's burgundy-black blood up through the thin hose. He loves butterflies. Without one, each vial has to be jammed directly into the base of the needle, which tends to rip up the vein. Butterflies are expensive, so not all labs use them.

Every piece of equipment Rosemary employs has evolved from basic bloodletting tools. The pressurized vials, which look like test tubes with color-coded caps, are a counterpart to bleeding bowls, large clay or pewter basins placed below the incision site to catch the blood. These were often graduated like measuring cups so the phlebotomist could tally the amount removed before discarding it. The modern syringe has a mixed heritage: Its housing is descended from the small glass cups used for suctioning blood from tiny cuts made in the skin. "Cupping" has a history almost as long as bloodletting. In practice these cups, heated over a flame, were applied to different parts of the body; a partial vacuum held them in place. Doctors used cupping for localized pain or if a patient was too young or weak to be bled properly from a vein. The syringe needle actually has the most ancient origins, reaching all the way back to the earliest human's use of a thorn or animal tooth to break the skin. Jumping forward to the early eighteenth century, the preferred implement for piercing a vein was the new spring lancet, as compact as nail clippers, with a trigger-activated blade. One Baltimore bloodletter so adored his spring lancet that he was driven to poetry. "I love thee, bloodstain'd, faithful friend!" one stanza began.

The most cringe-inducing tool of the bloodletter was the leech, although nothing in Rosemary's work space is remotely related, thank God. Like cupping, leeches had been used since antiquity as an auxiliary to venesection. Placed on the skin, these bloodsucking creatures, close kin to earthworms, fed on a patient until sated. After about an hour, they'd drop off. A doctor would typically employ five to ten at a time, although to be covered with fifty wasn't unheard of. Leeches were handy for hard-to-reach spots, such as up the anus, down the throat, or inside the vagina. Tiny thread leashes kept the leeches from getting lost. The leech of choice was the European *Hirudo medicinalis,* exported worldwide from Sweden and Germany. In 1833 alone France imported 41.5 million of the suckers. A standard part of medical practice, leeches were kept close by in water-filled clay or glass jars.

One would think that the huge gains made in understanding human biology from the Renaissance forward would've curtailed the popularity of bloodletting. But no. In fact, the practice

A woman self-medicating with leeches, as depicted in a seventeenth-century woodcut

reached its height in the eighteenth and nineteenth centuries. The Western world's most powerful people, receiving what was considered the very best of care, were needlessly bled, cupped, and leeched. Retired president George Washington's death in December 1799 was hastened by excessive bloodletting, for example, historians conclude. The president, sixty-seven and suffering from a severe throat inflammation, was tended by three top physicians who could have saved their patient's life had they had access to two things not yet invented: antibiotics and steroids. Instead, they bled Washington four times within a twelve-hour period, a total of 2.5 quarts. He died that day. It sounds like manslaughter to me, but the doctors' actions were considered both medically appropriate for the time and even heroic. Under less grave circumstances, the rule of thumb for a single letting session was to keep the vein open until the patient passed out. "Bleeding to syncope," this was called. In a statement of near Galenic aplomb, the English physician Marshall Hall wrote in 1830: "As long as bloodletting is required, it can be borne; and as long as it can be borne, it is required." Dr. Hall and his fellow physicians were, of course, a few facts shy of our modern understanding. A healthy person can, in fact, replace a lost pint of blood in about an hour, but it takes weeks for the oxygen-carrying red blood cells to return to normal levels. Thus, frequent and copious lettings served only to create in patients an endless cycle of chronic anemia. These days the amount of blood a phlebotomist withdraws for testing is minor, about half an ounce per vial. And rarely more than seven vials are collected. If a patient does faint, more than likely it's from a touch of hemophobia, fear of the sight of blood and/or needles.

Rosemary became a licensed phlebotomist in 1965 and thinks of herself as an old-timer in the field. "I started in a little mom-and-pop lab across the hall," she recalls, gesturing the direction with a tilt of her head. "In those days I'd both take blood and perform basic tests. Everything was done manually— cholesterol levels, blood sugar, enzymes, pregnancy tests." Wistfulness is just a flash in her eyes. Of course, she explains, the whole field changed virtually overnight with AIDS. Safety procedures tightened. New tests were introduced, others replaced, most now performed with computers at a facility across the Bay. The patients changed, too. For the past fifteen years the majority of Rosemary's clients have been gay men, like Steve.

"You must have lost a lot of patients," I say quietly.

Without glancing up from her work, Rosemary considers this. "Oh, gee," she begins, then changes course. "I've gotten to *know* a lot of patients," she replies with a smile to Steve.

"Okay, you're about done for today," she adds, watching the last vial fill.

Rosemary withdraws the needle while pressing down with a wad of cotton so large it looks like a chunk of pillow. "Hold that, will you?" she tells him. She deposits the butterfly and tubing into a Sharps container, a receptacle for used needles, and then affixes the clump of cotton to the site with half a foot of tape. And that's that. The whole procedure has taken no more than five minutes. I back out of the cubicle, Steve rises, and, while the three of us small-talk for a moment, I am struck by this tableau: Rosemary stands between us cradling in her hands the vials of his blood. *That's a part of his body,* I think; *it has passed through his heart.* Those five finger-shaped vials must still be warm, like she's holding his hand in hers. We say goodbye as she gingerly places each one inside a shipping container emblazoned with BIOHAZARD signs.

THREE

Biohazard

PRIOR TO IDL, STEVE HAD REGULARLY HAD HIS BLOOD drawn at a SmithKline Beecham lab near his doctor's old office. He stopped using this lab in 1994 and we didn't give it another thought until one day five years ago when I brought in the mail, which included a special-delivery packet from the blood lab.

"Nothing good ever comes by certified mail," Steve muttered, frowning, as he tore open the manila envelope and pulled out a letter from SmithKline Beecham's president, dated May 27, 1999. According to the letter, a phlebotomist who'd worked at the lab Steve frequented had reused needles from blood draws (butterfly needles, it turned out). The woman had admitted to doing this "occasionally," thereby possibly exposing uninfected patients to HIV, hepatitis, and other illnesses. (It was unclear whether her actions were intentionally criminal or inexplicably ignorant, but a year later she would be indicted on multiple felony charges of assault with a deadly weapon—dirty needles.) Records showed that Steve might have been one of her patients, the letter suggested; she wasn't named, so Steve wasn't sure himself. Those who wished to get tested for possible exposure could do so and receive counseling at SmithKline Beecham's expense.

While the letter was addressed to Steve, it wasn't written *to* patients such as him, I noticed. It never mentioned, for instance, that the phlebotomist, by reusing needles, could have exposed HIV-positive patients to mutated strains of the virus. It is not just HIV that can be passed on, but an infected person's entire drug-resistance history. Through reinfection, a patient already low on treatment options could be left with none. Steve put the letter aside and dug up his SmithKline Beecham records, finding he had used the lab eighteen different times.

The investigation of the phlebotomist became a sensational local news story. Reports focused chiefly on the possibility that uninfected patients had been exposed to HIV and hepatitis, which was neither inappropriate nor surprising. The accused had had contact with more than twelve thousand people over a period of many years, so the pool of potential victims was sizable. Even if they weren't infected, there would be grounds for lawsuits for their emotional distress. But Steve helped me see a perspective never addressed in media accounts—that of a man whose blood could have been the source of infection for an-

other, or even for many others, making him feel like an accessory
to crimes he'd been powerless to stop.

"The idea of someone treating my blood so carelessly . . . ," he
said to me, pausing to steady his words. His eyes narrowed. "The
possibility of my infecting someone else is horrifying."

I've seen that look in Steve's eyes one time since, late on a
Saturday morning. A few moments earlier, I'd told him to sit
down at our kitchen table. "Honey, I've gotta tell you something,"
I'd said shakily. "It's really important."

A lifelong insomniac, I'd been struggling through an awful
week of sleeplessness, as Steve knew. I'd rolled out of bed early
that morning feeling exhausted yet again. I took a look at my face
in the bathroom mirror. My red-rimmed eyes were so bloodshot,
I imagined they'd sucked my veins dry, left me iron-poor, a bit
anemic. So, I reasoned sleepily, I would be a patient of my own
Dr. Feelgood: I would give myself an injection of Steve's B_{12}.

With the exception of sleeping pills, I'd never once considered taking any of his medications—and there were a couple of
tempting ones, including Vicodin, which he used to treat his
neuropathy pain. But this was different: a little instant vigor. *It's
just a vitamin shot*, I told myself as I reached for the vial in our
kitchen cabinet.

The B_{12} injections were a new therapy for Steve. At his doctor's suggestion, he had been using over-the-counter oral B_{12}
supplements together with a B_{12} nasal gel, intended mainly as
an antidote to drug-induced fatigue. He'd found the nasal gel
messy, though, felt no less tired, and wondered, understandably,
how much of the vitamin his body was even absorbing. His
doctor's solution: a prescription for full-strength B_{12}, a single
one-milliliter injection per week. We picked it up that same day
along with a year's supply of needles, a box of multiple pouches,

the syringes within loose like Halloween candy. Steve's doctor had taught me how to give him the shots, and I'd already done it several times, the last one just the day before. I was a natural, even a tad smug about it. I had no fear of needles, never have, a trait reinforced by the fact that Steve, unflappable in most things, was creeped out by them. He couldn't even look when I gave him his shot.

In the semidarkness I grabbed one of the syringes, popped off the cap, and jabbed the needle through the glass vial's gray rubber top. I felt weirdly proud of myself; I'd come up with a brilliant solution for being sleep-deprived. *This is going to make me feel so much better!* I pulled back the plunger and watched, ebullient, as the syringe filled with the bright red medicine, foaming at the top like a glass of Strawberry Crush. I tapped out the bubbles. Oops, I'd forgotten the rubbing alcohol. I set the syringe on the countertop. Returning from the bathroom, I decided not to give myself a shot in the arm, as I would Steve, but in my butt, so he wouldn't notice. I wanted to hide it from him.

I pulled down my sweats, swiped a soaked cotton ball on my right cheek, pushed the needle in, pressed the plunger, and just as quickly pulled it out. There: a dewdrop of dark red blood, visual proof that the injection had dived through my white skin. I could almost see fizzy particles of B_{12} swimming to my heart, my eyes, my limbs, revitalizing me. I smiled at the imagined bursts of energy that would take me through the long day ahead. I slapped on a small bandage, restored the cap to the syringe, turned on the overhead light, and opened the kitchen cabinet.

And that's when I looked closely at the box of needles. Inside were two open pouches, one with new syringes, one with used. I had reached in blindly, grabbing the first needle I'd felt—a dirty one, I was now sure.

Already, I pictured, a speck of Steve's blood had entered my circulatory system. I shivered uncontrollably as it raced through my veins, pumped through my heart, seeped into my lungs, swept into my arteries, all the while multiplying, infecting every cell, flooding my body with HIV. What rose from the pit of my stomach and caught in my throat was not bile but blood, thick and sour. It tasted like fear.

I held my breath, as if to choke off all emotion. The moment I exhaled, fear filled the room. Had Steve walked in at that second, rubbing the sleep from his eyes, he'd have been overcome, too. I was having a panic attack; heart thrumming, ears ringing, it took all my strength just to sink into a chair.

Adrenaline was not living up to its reputation. It wasn't the superhuman jolt I'd have expected—that surge that allows a mother to lift a crumpled car off her injured child or that burst of mental clarity that lays out the world like precise moves on a chessboard. The reality was far from the fantasy, the latter owing heavily to the late-1970s TV show *The Incredible Hulk,* a guilty pleasure when I was in college. The transformation from scrawny scientist David Banner into the green behemoth was ignited by overpowering emotion. ("Don't make me angry," actor Bill Bixby would say, more warning than threat; "you wouldn't like me when I'm angry.") Muscles bulged, pants ripped, the shirt shredded, but the Hulk's transformation was never complete till he smashed through a wall or two.

In my case, my energy imploded. Thoughts raced, getting nowhere. I'd latch on to one and it was irrational. *I should suck out the HIV at the injection site, but how do I get my mouth to my hip? I should down a mouthful of Steve's AIDS drugs,* I told myself next. *Wouldn't that stomp out the infection?* At some point my mind had ground down to nothing, and I was aware of my heartbeat shaking me awake from myself.

I thought perhaps calm could be restored by my going

through the motions of everyday normalcy—showering, eating breakfast, getting dressed. *Breathe,* I coached myself. *Breathe.* In the weird fugue state of the guilt-ridden, I watched Steve get up and go through the same rituals I had. But then I couldn't bear the pretense of ordinariness any longer. I told him I had something to tell him. I asked him to sit down.

Once I'd spilled all, Steve pushed back from the table, stood, and turned toward the kitchen cabinet. He didn't say a word or, at least, none that I heard. I watched him pull the box of needles from the cabinet, place it on the table, and silently begin counting. He looked up.

"Needles come in bags of ten," he said, cool and clear. "I opened this new bag yesterday and we used one for my shot. If you'd taken a used needle, I'd be able to find only nine in the bag."

My mind was fuzz.

Steve was talking to me. I heard "ate." *Ate?*

"There are eight in the bag, Bill. See?"

I was starting to understand, coming to ground.

"You used a new needle—you're fine. Thank God. They're all here, except the one you took. Where'd you put it?"

Now fear gave way to shame, burning my face. "In the garbage."

"The garbage?" Steve pulled back, appraising me. "Oh, the trash man'll love that."

Steve went silent, clearly expecting me to speak.

"I'm so sorry," I responded. "I—I'm an idiot. I didn't sleep. I—"

"Look, I'm glad you're all right. But . . . do I have to hide these?" He tapped the box.

I didn't reply. I was still checking Steve's math in my head. "Eight in the bag? Are you sure that's right?"

He walked away from the table. "Count 'em yourself."

I counted. And did so again, later, while Steve went to Walgreens to get a Sharps container. Still, it took me a couple of days to shake the conviction that I'd used a dirty needle, that I might be HIV-infected. Likewise, it took Steve as long to understand my reaction, why my fear lingered. I felt as if I'd nearly been in a car crash, I was finally able to explain to him. And even though I knew I was perfectly safe, I could still hear the screech of tires, still feel the blood rush from the near hit, my pulse racing.

WHEN DECLARING A PERSON CLINICALLY DEAD, THE ATTENDING physician or EMT must note in writing that there is no pulse. The carotid artery, just below the curve of the jaw on either side of the neck, is the site most often felt. The pulse is both the last and, in the living, often the first of the vital signs checked. It is the heartbeat by proxy, each throb caused by the powerful contractions that propel oxygenated blood out into the arteries. This outward surge of blood carries such force that the vessels swell to accommodate it; hence a palpable, visible, sometimes even audible pulsation. In all, seven pairs of arterial pulse points dot the human body: at the neck, inner elbows, wrists, and both sides of the groin; in the pit of the knees; behind the ankles; and atop the feet. Typically, arteries are buried deep within the body, but at these points they lie near the skin's surface and over a bedding of bone. This makes them ideally situated for palpation, examination by touch.

In American Sign Language, the sign for "doctor" is the finger-spelled letter *d* tapped inside the wrist, which captures in a simple gesture the most fundamental part of a medical exam, the iconic act of taking the pulse. Performed in every culture, this basic diagnostic test is as old as the practice of healing itself. The careful placing of two or three fingers along a tiny stretch of artery used to be considered an art form, a notion largely lost in the mad shuffle of contemporary health care. One needs to page back a good hundred years or so to rediscover a time when this vital sign retained all of its, well, vitalness. I've found no more erudite an advocate than Sir William Henry Broadbent, personal physician to Queen Victoria and author of a unique monograph, *The Pulse* (1890). In its pages Sir William is a spirited defender of what he calls "the educated finger." In an early passage he subjects the wrist and its pulse point to a curious clinical analysis, as if describing a patient with an odd case history. He is longwinded, but endearingly so: "At first sight it seems strange that the radial artery, which supplies [blood to] merely the structures of a part of the hand—a few small bones with their articulations, a few muscles and tendons, the skin and nerves distributed to it—should afford the varied and far-reaching knowledge we look for in the pulse. The hand is not essential to life, it contains no organ of any importance, and *a priori* it might have been supposed that the variations in the circulation of the blood in so small a member could have no significance." There is little about this passage I do not love, from the doctor's crisp visual dissection to the delicious irony he's blinded to in his academic fervor: If not for the irrelevant hand, he could not even take the pulse, let alone write about it. But I digress. The distinguished doctor, who'd practiced medicine for more than thirty years by the time his book was published, goes on to state without equivocation that the wrist pulse is a "trustworthy index," a reliable gauge for the entire circulatory system.

A portrait reproduced on the frontispiece of his memoirs broadens my sense of the man: Seated, he looks the very essence of "bedside manner"—compassionate, patient—as though he's just asked, "What seems to be the problem?" A stout gentleman in his late sixties, I'd guess, the doctor is dressed in a dark formal suit with a wide satin cravat. A pocket watch is comfortably secreted in his closed palm. Perhaps he can feel the tick of its clockwork against his skin.

In his day the pulse opened a personal dialogue with the body, and a skilled clinician could glean an astonishing array of insights, far beyond a tally of heartbeats per minute. With nothing but his fingertips, Broadbent claimed he could assess the condition and health of the arteries, calculate blood pressure, and discern the emotional well-being or physical ailments of a patient. Even a person with profound sleeplessness was implicated by his pulse. The insomniacal artery, Broadbent wrote, was "full between the beats" and could be "rolled under the finger," while the pulse waves themselves ended abruptly, as if exhausted from the effort.

An impetus for writing his treatise was Broadbent's grave concern that physicians' tactile skills were eroding (or, among young doctors, not fully maturing) as technology was relied upon more and more. Back in the late 1850s, when he'd begun his lengthy career at London's St. Mary's Hospital, a newfangled device had started attracting notice, the "sphygmograph," an ingenious though initially clunky contraption that could create an ink tracing of a patient's pulse. It worked this way: With the wrist upturned, the forearm was immobilized. A small sensor plate rested atop the pulse point and, in essence, rode the gentle waves; the motion was translated simultaneously onto a strip of paper, forming a steady sequence of squiggles. To the medical community, an instrument that could provide an objective reading of the pulse was an important advance (the modern blood pressure cuff is the sphygmograph's direct descendant). However, while Broadbent used various models throughout his career, he was never a full convert. In *The Pulse* he praised the machine's ability to mimic what a skilled physician could do by hand but emphasized that it was "not an infallible court of appeal." The device was tricky to operate—it wasn't like placing a thermometer under a tongue. In fact, Broadbent maintained, many of the "niceties of information" were out of its reach; no machine could ever replace the power of human touch.

In mastering the language of the pulse, Broadbent was linked to a timeless tradition, one transcending cultures and medical philosophies. The physician-priests loyal to the lion-headed Egyptian goddess Sekhmet relied on pulse palpation to reach their diagnoses, as evidenced by tomb inscriptions circa 2000 B.C., and medical papyri from this same era contain repeated reference to the pulse. "The heart speaks out of the vessels of every limb," one particularly lovely line translates. In the history of medicine, however, the literature of ancient China is unmatched in its extravagant attention to deciphering the body's rhythmic code.

The Chinese text called *Huang Ti Nei Ching Su Wen* (*The Yellow Emperor's Classic of Internal Medicine*) is one of the world's earliest and most famous medical guides. Although the work is attributed to the legendary first ancestor of the Chinese nation, historians concede that it is the product of neither a single writer nor single time period but rather a compilation of many teachings over hundreds of years. The oldest portions may date as far back as the fifth century B.C. To me *The Yellow Emperor's Classic* is best appreciated not for its physiological accuracy but for its richness of ideas. All of the disciplines of traditional Chinese medicine sprang from its theories.

The entirety of what Broadbent could read at the wrist pulse was just the starting point for what *The Yellow Emperor's Classic* describes. By applying varying pressure to different points along that single stretch of artery, an accomplished physician could derive a full accounting of every internal organ as well as a sense of the subtlest qualities of yin and yang, the positive and negative cosmic forces that balance in good health. The physician intuitively correlated into his reading a bewildering string of external factors—the climate, the direction of the wind, colors, odors, tastes, sounds, the natural elements, the positions of constellations, and more—and arrived at a diagnosis. To a Westerner such as myself, this ability seems almost supernatural and farfetched. I find a stronger resonance in the text's evocative characterizations. The resting pulse rate of a healthy heart will resemble "a piece of wood floating on water," for instance, and the throb of a vigorous heart "should feel like continuous hammer blows." The pulses relating to unhealthy conditions are also lyrical. A sickly pulse might reverberate like "the notes of a string instrument" or feel like "fish gliding through waves"— descriptions that nonetheless thrum and flicker with life.

Dr. Broadbent was never so poetic. On the contrary, he encouraged physicians to express no personal style whatsoever when writing about a patient's pulse, thus eliminating the risk of ambiguity. The rate of pulse beats should be described as either *frequent* or *infrequent,* he insisted, with no shades in between. Arteries were *large* or *small,* and the "tension" or blood pressure within them *high* or *low.* What's interesting is that this colorless vocabulary obviously did not reflect his wonder at the pulse. "It is impossible to examine a large number of pulses," he enthused, "without being struck by the extraordinary diversity of frequency, size, character, tension, and force met with." Of course, Broadbent's contribution to his field went beyond the crafting of a glossary. During his nearly four decades at St. Mary's Hospital, he was able to confirm definitively the link between high blood pressure and disease, paying particular note to hypertension in late-stage kidney disease. He was also among the first researchers to elucidate the risks of low blood pressure. In his midsixties William Henry Broadbent was recognized as one of Great Britain's leading clinicians.

A year after *The Pulse* was published, he was contacted by officials at Buckingham Palace. The queen's grandson, Prince George of Wales, had come down with typhoid fever, and the doctor's expertise was requested. He remained in attendance at the prince's residence for a month, seeing the twenty-six-year-old through to a complete recovery. Not a week had passed before he was summoned yet again. Now one of George's brothers had been stricken by influenza, and he died in a matter of days. Word reached Dr. Broadbent that Her Majesty, Queen Victoria herself, wished to see him. Somehow I doubt he was expecting a promotion.

"The Queen sent for me about 3, and I had to tell her the whole story of the illness," he wrote in a letter to his sister, dated January 17, 1892. "She was sitting in an ordinary chair at a writing table, and of course I had to stand. I was there almost exactly an hour and a quarter." Though he betrays not a whit of emotion in this retelling, the royal visit did go well. Soon thereafter he was appointed Physician Extraordinary to Queen Victoria.

WILLIAM BROADBENT HELD THE QUEEN'S WRIST. NOW A QUEEN holds mine: one named Ernesto, the physician's assistant in my doctor's office. Up to this point on a recent visit, nothing extraordinary has occurred. A good forty minutes after my arrival, Ernesto flung open the inner office door and sang out, *"Willlllllyaaaammm!"* He then weighed me in the hallway, led me into a stuffy cubicle, quizzed me about why I'd come, and just as I was beginning to regret making the appointment—to broach the topic of anti-anxiety medication, no less—something relatively pleasant happened: The room went quiet. It was time for Ernesto to check my pulse.

At that moment it seemed as if a tiny Dr. Broadbent perched atop Ernesto's hooped earring, whispering instructions in his ear: "Three fingers should be placed on the artery, and it will not be amiss to observe the old-fashioned rule of letting the index finger always be nearest to the heart; the different points with regard to the pulse should then be ascertained, each by a distinct and separate act of attention."

Ernesto's technique is flawless: his grip, firm yet gentle; his bare hand warm. Utterly focused, he studies his wristwatch. He stands so close to me, I can feel his pillowy belly at my arm. I have the sensation of being anchored by this heavyset man as he listens with his fingers to my heart. I stop thinking about what brought me here and what Dr. Knox might say. For thirty seconds I am absolutely grounded in present tense.

Then Ernesto looks up from his watch, releases his grip. "Sixty-eight. Heart rate's sixty-eight," he says. "Perfectly normal."

At which point I feel tempted to compliment him back: *How fashion-forward of you to be wearing white clogs*, for instance. But no, I could never say that with a straight face. Anyhow, the moment is lost. He's already jammed a thermometer into a plastic sleeve and has placed it under my tongue. My pulse appears on my chart as a scribbled number at which Dr. Knox will scarcely glance. Dr. Broadbent would've been disappointed.

Today pulse palpation is a central part of an exam only in cases of serious cardiovascular disease. What's more, in hospitals and many doctors' offices, beats-per-minute is often obtained not by hand but through a monitor attached to the blood pressure cuff or a sensor clipped like a clothespin to the index finger. These digital devices, sensitive enough to detect the heart rate through capillaries in the skin, operate just like the pulse calculators built into sports watches, stationary bikes, and so forth. They're used in the interest of speed, accuracy, convenience, and, I'm told, patient comfort. Some people do not like to be touched. While there's no such high-tech revolution yet under way at Dr. Knox's office, Steve's doctor visits are different. In an office gone digital, the haste with which his body is stripped of its secrets—weight, body temperature, heart rate—is dizzying. Every second shaved from an exam is, of course, money saved by an HMO. But at a time when patients are encouraged to turn to WebMD with the questions their family doctor didn't have time to answer, it strikes me that pulse taking by hand remains an uncorrupted tradition, one with strong roots in the classical age.

In ancient Greece the art of feeling the pulse (*sphygmopalpation,* from the Greek *sphygmos,* for "throb") was first taught by the physician Praxagoras, a contemporary of Hippocrates, one of the earliest fathers of Western medicine. Praxagoras's star pupil, Herophilus (335–280 B.C.), was the first physician to methodically time the pulse. He used a primitive water clock that had been invented to time the speeches of orators. Erasistratus, Galen's phantom bloodletting rival, is credited with incorporating the pulse into clinical exams. His first diagnosis: lovesickness, in a young man whose pulse quickened dangerously whenever his crush drew near. The attention paid the pulse at that time is all the more impressive given that the ancients were missing huge pieces of the puzzle. Though these healers knew they had their fingers on the pulse of the body's innermost workings, they did not understand the actual role of the heart in circulating blood any better than they knew the distinction between veins and arteries.

Not until the intellectual watershed of the Renaissance did this begin to change. A major upheaval in how the body was viewed required first the systematic dismantling of the hallowed teachings of Galen. A key figure in this deconstruction was the Belgian anatomist Andreas Vesalius, who, in his illustrated seven-volume masterwork of 1543, soundly disproved two hundred of Galen's factual errors. No, the liver did not distribute blood throughout the body. No, blood did not "sweat" from the right side of the heart to the left. No, animal anatomy wasn't interchangeable with human. And on and on. Vesalius, among others, paved the way for Great Britain's William Harvey, who in 1628 turned the world on its ear: Blood circulates, he announced in his historic *An Anatomical Essay on the Movement of the Heart and Blood in Animals*. For its role in launching the modern era of medicine, contemporary historians have called Harvey's book one of the three greatest works in the English language—all three, curiously, dating from the early 1600s— alongside the King James version of the Bible (1611) and the First Folio Edition of Shakespeare's plays (1623). By comparison with these other two works, Harvey's tour de force is small in size (five by seven inches), short in length (seventy-two pages), and written in deceptively simple language.

"The movement of the blood in a circle is caused by the beat of the heart," he declared, summing up in one sentence his entire theory of the circulatory system. Then, as if to head off any *But what about . . . ?* from the unconvinced, Harvey added, "This is the only reason for the motion and beat of the heart."

Through animal vivisection, human dissections, and observations of living patients, Harvey poked more holes in Galenism. Blood did not ebb and flow within the same vessels, as the Greek physician had taught. Instead, the arteries carry it away from the heart, and the veins bring it back. Valves inside the veins help the depleted blood make the return trip. Further, although he couldn't explain how, Harvey theorized that blood passes via some unknown mechanism from the arteries into the veins. The crude new microscopes of his day were not nearly powerful enough to reveal the minute bridging vessels now known as capillaries. In a final slap to Galen, Harvey also proved that the arteries themselves did not contract and dilate like blacksmith bellows, thereby producing the pulse. "The pulsation of the arteries," Harvey wrote, "is nothing else than the impulse of the blood within."

Accomplishments notwithstanding, Harvey was not necessarily a "better" scientist than Galen, contemporary writer-physician Jonathan Miller contends. "The difference between the two men is not one of ingenuity and skill—in fact, if these were the sufficient conditions of scientific progress, Galen rather than Harvey might have been the discoverer of the circulation of the blood." Instead, the difference between them was one of

William Harvey

"metaphorical equipment," Miller argues in his book *The Body in Question* (1978). Galen likened the heart to a common household item of his time, the oil lamp: The organ heated and transformed blood from a dusky fuel to a flaming scarlet stream, illuminated by Vital Spirits. In his reckoning, however, that was the extent of the heart's role. "Galen's inability to see the heart as a pump was due to the fact that such machines did not become a significant part of the cultural scene until long after his death," Miller states. By the end of the sixteenth century, though, mechanical pumps began to be widely employed in mining, firefighting, and civil engineering, such as in the design of ornamental public fountains. Therefore, when Harvey conducted his experiments (among them, watching as hearts slowly failed during animal vivisections), he was able to see the organ for what it was: a pump, resembling the marvelous inventions in use around him.

With the medical community electrified by Harvey's discovery, a new interest was sparked in injecting substances directly into the bloodstream. But a simple means for such a procedure did not exist. Enter: British architect Christopher Wren. In 1656 Wren fashioned a crude syringe from a hollow feather quill fastened to a bladder and was able to pump opium straight into a dog's vein, thus creating a method for IV therapy as well as one very mellow pooch. Wren's success inspired others to infuse animals with not just medications but also wine, beer, milk, urine, anything liquid—often with fatal results—and eventually to try blood. In 1665 the British anatomist Richard Lower performed the first successful blood transfusion in animals, linking one dog's artery to a recipient dog's vein with a quill piping. The transfused dog had first been bled almost to death, so its fast return to vim was hugely dramatic, bordering on miraculous. The floodgates were now thrown wide open.

The next step: animal-to-human blood transfusions. Over the following couple of years, a spate of attempts were made, though none for what we'd now consider logical or medically appropriate reasons—to treat hemorrhage, say, or to bolster red blood cells in acute anemia. The physiological unknowns at the time were considerable. Neither the component parts of blood nor its role in transporting oxygen, nutrients, hormones, and pathogens had yet been discovered. Interestingly, the idea of blood compatibility *was* considered, though not in the modern sense of the phrase. (Blood typing did not arise until the early 1900s.) Rather, a transfuser had to be cautious when mixing blood because it contained qualities. As perfume was the essence of a flower, so blood was a concentrate of traits, whether in man or beast. A fearless soldier had brave blood, for instance. A raging bull had angry blood. In theory, then, a transfusion had the potential to restore strength to the weak, calm to the crazy,

Emblematic representations of the four temperaments associated with each of the humors of the body. A slight excess of one humor determined whether your natural disposition was sanguine (surplus blood), choleric (yellow bile), phlegmatic (phlegm), or melancholic (black bile). Engravings by sixteenth-century German artist Virgil Solis

and so on. Hence in 1667 French scientist Jean-Baptiste Denis introduced the docile blood of a calf into the circulatory system of a raving madman. But did it work? Well, Denis thought he had triumphed. The recipient had vomited profusely and urinated what looked like liquid coal—he was being purged of his lunacy! From a modern take, however, we know the man was suffering a severe transfusion reaction and was lucky to have survived. But the story didn't end there. Before a follow-up transfusion could be performed, tragedy intervened. The man's long-suffering wife had finally had enough and administered a lethal dose of arsenic, thus bringing both the marriage and the experiment to a close.

News of Denis's initial "success" emboldened scientists to consider human-to-human blood transfusions. To the great minds of the seventeenth century, William Harvey included, this seemed like sound science because a belief in humoral theory was still widespread. A person in good health always had slightly more of one humor than the other three, and this excess determined the kind of person you were. Extra yellow bile made you *choleric*—a disagreeable sort. A tad more blood and you were *sanguine*—cheerful, optimistic. Remnants of this Doctrine of Temperaments, as it was known, survive to this day in the related words *melancholic* and *phlegmatic*. In an extrapolation of these factors, a German surgeon named Johann Elsholtz proposed in 1667 the use of transfusion as a remedy for marital discord. Would not the mood of a melancholic husband be lightened by transfusing him with the blood of his effusive and sanguine wife? And, flowing the other way, might not the wife become more temperate? The mutual exchange of blood between mates could heighten understanding between them—seventeenth-century couples therapy without all the talking.

Elsholtz never had the chance to move beyond the hypothetical, however. Magistrates throughout Europe could not ignore the reality that transfusions were killing people, and a ban was implemented in 1668. (In fact, it would be another 250 years before safe, effective human-to-human transfusions would be performed.) Though relegated to a minor historical footnote, Elsholtz was nevertheless on to something, I choose to believe, if only by a shiny thread of whimsy.

. . . .

WITH THE POTENTIAL FOR DISEASE FACTORED OUT, TO BE INFUSED with what runs in Steve's veins would mean being imbued with, among other qualities, his innate sanguineness and his long-lived love of comic books. The latter started in the summer of 1975 at Lefti's corner store in East Hanover, New Jersey, where he grew up. Steve was twelve when he picked up an issue of *Fantastic Four*. It was about a family, he thought, albeit an unconventional one—three guys and a girl, two of them related by blood, united in fighting on the side of good. Steve, one of four kids himself, found it fun, but another title in the Marvel Comics Universe really grabbed him: *X-Men*. With his first issue, *Giant Size X-Men* #1, he was hooked. That it was a number one played a part. Like everyone else in his family, Steve was a collector—Wacky Pack gum cards and Flintstones jelly glasses were favorites. Now he had the starting point for a new collection, one that would grow over the years to thousands of issues currently stored in long boxes in all our closets.

From its inception, what made *X-Men* different from other comics was that it introduced the idea of mutants into the superhero pantheon. These characters weren't the victims of freakish science experiments gone wrong or of sudden exposure to mysterious biohazards; they were born that way—they had a genetic quirk in their DNA, an X-factor. Their powers, though, often didn't manifest until their teen years. You woke up one morning to find your body was starting to change. The intended parallel was to puberty, but to any readers who saw in themselves something shameful, the X-Men struck a deeper chord. Although they were heroes doing good, the mutant X-Men were grossly misunderstood, despised by society at large, hunted down by the government. Where Superman was lauded in the bright light of day, the X-Men had to stick to the shadows.

With its monthly tales of prejudice and perseverance, the comic book slowly instilled in Steve a resolve that would make his coming out far less torturous than mine. Not being an only son or raised a devout Catholic also helped ease his way. It seemed perfectly normal to him to keep secret his "identity" while at the same time accepting it as a natural part of himself. Just as mutations occurred in nature, so did homosexuality. He also knew a time and a place would come when he could safely expose this aspect of himself. High school was just not it. He never doubted that a real-world correlate to the X-Men existed, a group somewhere who'd accept him.

I read the occasional comic book as a kid, yet they were never a constant in my life. While I could've named the major superheroes, my taste ran more to *Richie Rich* and *Archie's Pals & Gals.* Now, viewing superhero comics through Steve's eyes, I see not only how much they've evolved but also how, with their godlike heroes and grand-scale drama, they are like the ancient Greek tales I've always loved. Superhero comics are the medium of modern myths.

Their unique dynamism, I've learned, hinges on a device that's crucial to this art form: the blank space between panels— the gutter, it's called. Much happens in these narrow strips of nothing. There, your mind takes two scenes and bridges them, filling in the elements that are not drawn or lettered. A fist is thrown in one panel; the villain careens backward in the next; but you envision the wallop. The moment of impact and the crunch of cartilage are your creations, as is the breadth of emotion. An eerie calm can stretch as long as you decide. This involvement turns you from a mere reader of the comic book into a collaborator, a member of the creative team that makes the story work.

Time passes at a slower rate in a comic-book universe than in our own. While it's been almost thirty years since Steve picked up his first superhero comic, in Marvel Time, as it's called, only a few years have gone by. So when Steve reads the latest issue of *Uncanny X-Men,* say, he meets up with old friends who've hardly aged since he was a kid. Up till now I'd thought this was the whole appeal, a sweet nostalgia trip for a forty-year-old man battling AIDS, a well-deserved escape from his reality. But it's clearly more. There's a powerful draw in a stack of comics. In their pages, overwhelming odds are overcome. Good guys win. Death is not always final. And the question *What comes next?* is never frightening. It's exciting.

The only time he reads comic books, I notice, is at bedtime. It's shortly after he's taken his handful of nighttime meds. His stomach roils, sorting out the pills, sending them out through his blood. His feet, pinging like sonar from the pain of his neuropathy, kick at the sheets. His fingers are almost too numb to turn the thin paper. Though the sedating effect of the drugs sets in, he fights to stay awake, to read another page, then another, just one more. I give up before he does and turn out my reading light. Before drifting off, I look over. Steve's smiling. He's lost to another world, fighting the good fight in the space between panels.

FOUR

Blood Sister

WHEN I WAS A LITTLE KID GROWING UP IN 1960S Spokane, I associated blood with the rough-and-tumble world of brothers. Though I had no brothers of my own, I could always go to my best friend's house to be among some. Conversely, Chris Porter came over to mine to be around sisters, for he had just one and I had a surplus—five. I almost never saw blood at our house. My sisters played board games, not ball games. Twister was about as rough as it got. Sure, we had Mercurochrome and a tin of Band-Aids in our medicine cabinet for skinned knees and mosquito bites so scratched over they bled. The Porters, by comparison, had an actual first-aid kit, stocked with pads of gauze the size of sandwich bread, splints, and a tourniquet. *A tourniquet! How cool was that?* Their house was a two-minute bike ride away, a place expressly outfitted, I now realize, for boys to burn off energy. Outside, Chris and his three brothers had a basketball hoop mounted in a cement-filled tire, a tree fort, and a garage filled with every sort of sports weaponry imaginable—lawn darts, baseball bats, and cracked hockey sticks still good for whacking crab apples into the neighbor's yard. In the downstairs rec room there was a pool table and a punching bag and a floor so often cluttered with stuff—strips of Hot Wheels track, zillions of Matchbox cars, plastic soldiers, Erector Set buildings, Lincoln Log barricades—that Mrs. Porter routinely used one of those wide janitorial brooms to clear a path for herself to the pantry area, mercilessly toppling the mini metropolises in her way. She had a don't-mess-with-me severity my mother lacked, an I-don't-have-time-for-this quality, but the most radical difference between the two of them was that Mrs. Porter worked outside the home, something no other woman in our neighborhood did. She was the part-time nurse to her husband, Dr. Porter, a GP with an office nearby. Though always back home by the time school let out, Nurse Porter was never off duty. She knew back then, for instance, that I had what would now be called "white coat syndrome"—the skyrocketing of blood pressure and anxiety during a doctor's appointment. I liked Mr. Porter, but *Dr.* Porter terrified me. To get around this, during a lull in Chris's and my playing, she would sweep in, strap the blood pressure cuff on my arm, and, before my heart could start racing, she'd have already pumped and squeezed out the result. "See?" she'd say to me. "Perfectly normal." Oh, she was crafty, that Mrs. P. And unflappable. I remember once being out under the carport with

Chris when little Melissa Parker ran up wailing in a voice that could've shaken the fort from the tree: "Andy cracked his head open!" Sure enough, her bloodied brother, wheelbarrowed by two friends, soon bounced up the driveway. The Porter boys and I watched, straddling that gulf between horror and fascination, as their mom calmly sprang into action. Alas, so much blood for what ended up needing so few stitches! Time and again, as spectator and sometime recipient, such injuries reinforced in me the same equation: Blood was a guy thing, not a girl thing. Little did I know that there was a tide of female blood in my own home, and it seldom ebbed.

As the only boy in an Irish Catholic family, I was deeply conscious of how differently my parents viewed a son as opposed to daughters. The fifth of six kids, from the earliest age I felt genuinely prized, an individual whereas my sisters were often lumped together. We were "Billy and the Girls," like a pop band in which, long before I could talk, I'd been named lead singer. The Hayes daughters were raised with the expectation that they'd eventually marry and have children. I was led to believe I'd go to West Point, as had Dad, carry on the family name, and one day take over the family business, a Coca-Cola bottling plant. Only-boy-ness also meant having no hand-me-downs, whether clothes or bicycles or books, plus exclusive access to Dad, who took me alone to the drive-through car wash and to "he-man movies" such as *True Grit*. As the supplier of soda pop to all of Spokane's sporting events, he received free passes to hockey games, boxing matches, the annual rodeo, and off we'd go. It was as if manliness were a destination to which Dad regularly led me. Father and son, we'd sit in the bleachers most Sunday afternoons, sharing bags of roasted peanuts and time away from "the squaws," as he called my sisters and mom. We'd make it home just in time for dinner. As it was every night, the dinner table was like a game of musical chairs, the girls constantly pop-ping up to fetch this or that while Dad and I remained seated, never lifting a finger.

I'd had my own bedroom since the summer after my seventh birthday. Before that, I'd roomed for as long as I could remem-ber with my sister Shannon, who was then unceremoniously moved in with "the baby," four-year-old Julia. Shannon was two years older than me and the sister to whom I was closest. To-getherness hadn't ended with our getting separate bedrooms. Her best friend, Mary Kay, was Chris's sister, so our paths often also crossed at the Porters', as well as at school and catechism class. Our connectedness as children was one of complements: Her emotions bubbled over, I held mine in. It's something we still joke and talk about today: Shannon cried enough for the two of us, if not the whole family. And yet, as the fourth daughter, she was always somewhat misplaced, not allied with the eldest three and rarely getting the attention from Mom and Dad both Julia and I received. Though younger than Shannon, I tried to act like her protective big brother.

To the senior daughters, Colleen, Ellen, and Maggie, I was the baby brother they doted on but who also got in their way every day in the tiny bathroom we all shared. We called it "the yellow bathroom," for it was tiled the dusty color of lemon drops. We never shared the bathroom to the extent of bathing or using the toilet in front of one another; the locked door guaranteed privacy. But in the final minutes before bolting out of the house for school or church, we all ended up in there at once. In the large mirror above the twin sinks, my sisters and I were a jumble of pressing bodies, a photo booth filled to capacity. From mem-ory, I pluck a typical scene: It is a school morning in 1969. I'm a second-grader at Comstock Elementary. We've all got to be out of the house in fifteen minutes.

I'm in there first, as my bedroom is right next door and I'm already dressed, having laid out my clothes the night before—brown cords, a white shirt, and a belt; not much to it. At either side of the sinks are three drawers labeled with our names on masking tape. Colleen, Ellen, and Maggie have the left side; Shannon, Julia, and I, the right. Based on actual need, I could use the tiny ledge behind the toothbrush holder for the few possessions my bathroom drawer houses. It rattles as I yank it open, the lonesome sound of a comb and a tie clip. By contrast, the girls' drawers barely close, containing overgrown thickets of hairstyling paraphernalia and such.

I load Crest onto my toothbrush as Colleen enters, the first sister in. The eldest, she's always the first at everything. First to be confirmed at St. Augustine's, to go on a diet, to part her hair in the middle, to enter high school. A freshman at Lewis and Clark, she is twice my age, which, in my reckoning, puts her in roughly the same age bracket as redwoods and our parents. Colleen wants to be a fashion model; she has the most to accomplish here. She starts by tugging from her dirty-blond hair the pink foam rollers she's slept in, tossing them one by one back into her drawer.

Ellen and Maggie are next in, and the four of us automatically reconfigure. I now sit on the toilet seat, ostensibly still brushing my teeth, as Colleen takes the middle position, and Ellen and Maggie commandeer the sinks. Though they're roommates, close in age, and both attend Sacajawea junior high, they are opposites. Ellen is most Dad-like in being the bossiest child, an excellent student, and a voracious reader. She also has terrible eyesight, her one apparent vulnerability. When she takes off her thick lenses to wash her face, she can hardly see well enough to find a towel. No-nonsense in every way, she reminds me of Velma, my favorite character from *Scooby-Doo*. Maggie, like

Mom, is artistic. Her wrists jangle with jewel-colored bracelets she's made herself from debristled, melted-down toothbrushes. Her fingernails might as well be painted with nail polish— they're stained scarlet with Rit fabric dye from her latest project, a batik bedspread. Maggie sidesteps the parental ban on wearing makeup by dabbing Vaseline on her lashes, then crimping them with a torturous-looking eyelash curler. She hides her Bonne Bell lip gloss in her purse, to be applied at school.

At this point there's not much room for Shannon, a fourth-grader at Comstock. With great hesitance, as if it's a tough decision, she selects from her drawer the brush she always uses and starts untangling her long, dark-chocolate-colored hair. Ellen, finished with rubber-banding her braces, steps behind Shannon and plucks the brush from her hand. "One braid or two?" she asks.

"One," Shannon says. "I don't wanna look like Pippi Long-stocking again."

In toddles Julia, who is just five and, like me, drawn more to the crowd than by any pressing bathroom business. As I rinse and spit, Julia's eyes follow me, expectant. With deliberate showiness, I take up a clean plastic cup, twirl in some Crest, then blast it under the faucet so that the cup bubbles over with fluoridated froth. Julia beams as I hand her the toothpaste float, and she promptly dips her nose in it. She soon has the bathroom to herself, however, as the older girls fly from the house to the honks of carpools, and Shannon and I scurry down the block to Comstock.

As much as I loved being around my sisters, as familiar as they were to me, I also found them mysterious at times, particularly the eldest three. Some mornings, for example, they mentioned things in the yellow bathroom I did not understand, lapsing into coded girltalk before shooing me from the room, locking the door behind me. From my bedroom, ear to the adjoining wall, I could never quite make out their muffled chatter. Of course the Hayes girls did in fact share a private language, and one afternoon, when I was ten, the secret decoder ring was placed in my hand.

Panicky and insistent, Shannon pulled me into the bathroom. She looked as if she'd committed some horrible sin and was expecting to hear at any moment the booming voice of our father. Normally, when either of us got into big trouble, we'd be each other's first confessor and consoler, so I asked, "Did you do something wrong?"

"Cramps," she said. "I started having cramps."

Had Shannon's and my entries into puberty coincided, I might have been quicker on the uptake. But I was a fourth-

The Hayes siblings in 1967, left to right: Julia in Mom's lap, me, Shannon, Maggie, Ellen with the white gloves, and Colleen

grader who'd yet to sprout a single body hair, have my sex talk with Dad, or see the infamous health education film reserved for fifth-graders.

Cramps, though, was not an unfamiliar word. It was how Ellen excused herself early from the dinner table, without even asking permission from Dad. It was Maggie's password to freedom from attending church, the excuse that was never denied. I, too, had had cramps, with a bellyache or the stomach flu, but boy-cramps were far less contagious than girl-cramps. No sooner did one sister begin to feel better than another was murmuring "Cramps" and disappearing behind the yellow bathroom door.

As Shannon pulled me down next to her on that cool tiled floor, she appeared far more upset than I'd ever seen any of my other sisters. She sat against the toilet, which seemed prudent. She looked like she was going to upchuck.

"So, are you sick?"

She took a long time to answer "No," which left me certain the answer was yes. "I got my period," Shannon sputtered.

As I later learned, Mom had anticipated this day. A few months earlier she had given Shannon a well-worn booklet with a daisy on the cover, titled *Now You're a Woman,* and had sent her into the yellow bathroom to read it behind the locked door. Once she'd finished, Shannon handed it back to Mom, who had joined her in the bathroom. Mom must've thought the booklet adequately addressed the essentials of womanhood; they did not further discuss what being a woman now meant, but how to conceal it. She showed Shannon how to put together a menstrual belt; gave her a can of FDS feminine deodorant spray; and instructed her to deposit spotted underwear or bedsheets directly into the washing machine, never to leave them in the bathroom hamper.

They should've made a daisy-covered booklet for boys, one with just enough answers to help a brother help a scared sister. As Shannon opened up to me, I felt as if I were scrambling to assemble a jigsaw puzzle without having the cover to the box. It was clear that Shannon was bleeding, that she would keep bleeding for an entire week, and that there was no way to stop it. No wonder she looked so fearful. I, too, was frightened. I was also sworn to secrecy. Had I gone to Mom or Dad, I'd have gotten Shannon into trouble. Our mother had given her two last instructions: "Don't tell Dad. And don't tell Bill."

ON SOME LEVEL I MUST'VE TRUSTED THAT MY MOTHER KNEW WHAT she was doing with Shannon, but at the time she just seemed mean. If told then what I know now, I'd have been far more rattled. From an anthropological point of view, my mom's exacting this final vow from Shannon was simply a modern instance of an age-old custom: secluding menstruating women and girls. In practice, secrecy is seclusion through silence, a wall built around a girl at menarche, her first period, that remains standing today, to some degree, in homes throughout the world. In a broader context, however, 1971 Spokane was a pinnacle of enlightenment compared with almost anywhere, say, a hundred years prior.

In the late nineteenth century, British social anthropologist James Frazer recorded various shocking instances of ritualized physical and social seclusion. In his book *The Golden Bough* (1890) he described, for example, the young women of the Kolosh Indian tribe in Alaska who, at menarche, were confined to an individual hut with but a tiny opening for fresh air and food. The secluded girl could drink only from the "wing-bone of a white-headed eagle," which at first sounds like a privilege—the kind of vessel, say, reserved for a tribal chief—but wasn't. So unclean was she rendered by her menses that the entire water supply had to be protected from her lips. She was kept in this hut for a whole year, Frazer explained, without sunlight, exercise, or a fire's warmth, attended to solely by her mother. The length of the seclusion spoke to the depth of her community's fears. With her first period, the most potent, a girl became a destructive influence that needed to be neutralized. As she was inseparable from the blood, both had to be separated from society. Her power was phenomenal. With a glance, she could spoil the hunt or strike men dead.

On an island in the Bismarck Archipelago in the southwest Pacific Ocean, girls were confined for up to five years in hanging cages to shield the ground from their polluting touch, according to Frazer. Upon getting her first period, a young woman of a tribe in southern Brazil was stitched into a hammock, leaving only a button-sized airhole, as if she were a butterfly shoved back into its cocoon. Keeping her in darkness was essential; she could poison the sun with a look. Similarly, the Native peoples of both southeastern Bolivia and British Guiana (now called Guyana) shrouded pubescent girls in pods hung from the rafters of darkened huts. Here they remained for months, "suspended between heaven and earth," Frazer lyrically observed.

It all seems too awful to be entirely true. And indeed, one must question whether Frazer was embroidering or maybe just misinformed. After all, the occasional curious little girl surely disproved tribal beliefs by harmlessly peeking at the sky, say, or at a kid brother. Sure enough, as historian George W. Stocking Jr. suggests in his introduction to the current edition of *The Golden Bough,* a degree of skepticism is merited. Frazer, a man whose aspirations were more literary than scientific, was an armchair anthropologist who received his field reports secondhand, if not third or fourth. Knowing this helps explain why, despite Frazer's liberal use of labels such as "savages" and "barbarians," his accounts have such an engaging fable-like quality; even the book's title sounds drawn from the Brothers Grimm. The stories of caged and suspended girls call to mind tales such as that of Rapunzel, kept for years in an impenetrable tower beginning at the pubescent age of twelve.

In contrast to Frazer's method of observing life at a remove, the French historian Jules Michelet (1798–1874) rolled up his sleeves, conducting studies of menstruation that were nearly as invasive as gynecological exams. Michelet, renowned to this day for his panoramic *Histoire de France,* kept a private diary (published posthumously) in which he recorded in graphic detail the menstrual cycles of his wife, Athenais, who was thirty years younger than him. Entries included subtle observations of her daily flow—color, volume, density, odor—as well as an analysis of his own feelings, not hers, about her bleeding. Despite this particular fascination, his view of women in general was no more enlightened than was typical for his time. He reiterated in his essay *"L'Amour"* (1859) the belief that menstruation was a mark of women's natural *"débilité mentale et physique,"* which sounds no less insulting in French.

His opinion was an echo of Aristotle, who, writing more than two millennia earlier, declared menstruation as proof of women's inferiority. Aristotle also saw in bleeding an almost supernatural component. A menstruating woman's reflection could stain any mirror with a bloody cloud, he stated in *De Insomniis.* Such superstitions can be found, tenfold, in the writings of the first-century Roman author Pliny the Elder. In his *Natural History,* a thirty-seven-volume encyclopedia that remained a credible scientific resource up through the Middle Ages, Pliny warned that the touch of a menstruating woman turned wine sour, made crops wither, dulled razors, rusted iron, killed bees, and caused a horrible smell to fill the air. "The Dead Sea, thick with salt, cannot be drawn asunder except by a thread soaked in the poisonous fluid of the menstruous blood," Pliny wrote. "A thread from an infected dress is sufficient." He was also certain that menstrual fluid could make a potent impact on natural events. If held up to flashes of lightning, for instance, it could halt a hailstorm or a whirlwind. But not a volcano, sad to say. Pliny died at Pompeii in A.D. 79 while studying the eruptions of Mount Vesuvius.

It's easier for me to understand viewing the Earth as the flat center of the universe than to fathom how such mistaken ideas of menstruating women endured. I have to wonder if Pliny, who lived well into his fifties, ever spent extensive time at home with a wife and daughters. Did the women in his life concur with his notions? More recent accounts, some written from an instantly more credible perspective—female—place menstruation in a broader social context. Among the customs of the Pacific Northwest's Spokane Indians, the original inhabitants of the region where I was raised, a girl at puberty was temporarily moved to her family's menstrual hut, a comfortable space where she was cared for by her mother, aunts, and grandmothers. This was hardly the cramped cage Frazer envisioned. The girl was welcomed into womanhood through intimate education on sexuality, health, tribal taboos, and social responsibilities. Though this tradition died out by the late nineteenth century, similar and even more progressive customs are observed today among Native Americans such as the Shoshoni of Nevada. Once a month, women retreat to separate quarters, leaving behind the men to take care of the kids, cooking, laundry, cleaning, and other chores. The men gain appreciation for the women, who in turn enjoy a week's respite, an arrangement, as social anthropologists note, that helps foster cooperation and healthy relationships within the tribe.

Unique rites also take place within the confines of family, passed on generation to generation. My friend Maurice, who grew up in a small Brooklyn apartment in the 1930s, remembers with a touch of awe the privileges his older sister enjoyed whenever she had her cycle. In this close-knit Jewish household she ordinarily shared a bedroom with Maurice and their brother Jack; on many nights all three even snuggled in one bed. But the room was hers alone when Natalie, nine years older than Maurice, got her period. She secluded herself behind a locked door while he and Jack got booted to the couch. Even more luxurious than being given her own bedroom, Natalie was allowed for the week to smoke cigarettes, an indulgence denied the boys. Maurice still recalls the scent of her Chesterfields wafting through the keyhole, her room, he imagined, filled with pillowy clouds.

This wistful scene plays like a sweet spin on a grim script from Leviticus 15: When a woman has her period, the Old Testament prescribes, "she shall be in her impurity for seven days, and whoever touches her shall be unclean. Everything upon which she lies during her impurity will be unclean." This idea found its way into our five-bedroom Spokane home through multiple routes. One started at the supermarket. On trips to Rosauer's grocery, Shannon, like all my sisters, became well acquainted with The Aisle, an isthmus of pastel-colored cartons, a place boys didn't go. Mom would usually send me off to get some cereal or to spin the comic-book rack while she and Shannon entered that female zone. There, the packaging was pale and the wording vague, as if the products were specifically designed *not* to be noticed.

Meeting up again at checkout, me hugging a king-sized box

of Frosted Flakes, I'd see in our cart the familiar lilac Kotex box and other items of "feminine protection," a word pairing of indeterminate promise. "Protection" from what? (Not prying eyes, I admit, although my unwrapping one of the mummified wads in the bathroom trash certainly discouraged a second airing.) Were I pressed into listing items of masculine protection, I'd have said a football helmet, catcher's mitt, sports cup—gear to shield guys from outside injury. But girls had to be protected from themselves, from their own bodies.

This notion may have also been brought home from church with an oft-heard passage from Genesis on the consequences of Original Sin. God, punishing Eve for tempting Adam with the apple, tells her that He will "greatly multiply thy pain." Though this is a reference to the pain of childbirth, biblical scholars contend that the meaning was deliberately misconstrued by early church fathers to include menstruation. Monthly pain was part of the punishment all women had to bear for Eve's sin, a notion popularized in a common euphemism. Hence the so-called curse of Eve became simply "the curse."

NOW, AT AGE FORTY-THREE, WHENEVER I HEAR THAT EXPRESSION "the curse" I think of Shannon, who was bedeviled by painful and heavy menstrual cycles throughout her teens and, in more recent years, by a series of gynecological health scares. The image that comes to mind is Henry Fuseli's moody Gothic painting *The Nightmare* (1782), in which a defenseless, nightgown-clad woman is splayed atop her bed, except that it's daytime in my version, Shannon's wide awake, and the demon perched on her abdomen looks as if it is hatching plans: What vex to inflict next? I can see Shannon in that painting at all different ages—as a frightened girl, as a lonely teen, and as a vulnerable young woman.

This whole picture changed two years back when Shannon had a partial hysterectomy, surgery her doctors had recommended due to recurring, unusually large fibroids on her uterus. To celebrate this major life change, she joyously threw decorum to the wind and held a "Uter-Out-of-Me Party" at her Seattle home a week prior to the procedure. I was disappointed not to be able to fly up from San Francisco to attend, though she filled me in on the details by phone later that day. It was all good silly fun, she and ten women friends raising flutes of champagne to Shannon's uterus and bidding good riddance to tampons, panty liners, diaphragms, and bleeding. A friend who'd had the same surgery a year before brought quiche and deviled eggs in honor of the ovaries Shannon's surgery would leave intact, and organized party games, including rounds of Operation, for which the board game's male patient was turned into a she with a felt pen.

Shannon sounded strong, happy, her voice fizzy with high spirits. Still, I, the worried brother, wondered if she was having any last-minute doubts about the surgery.

"No, I'm ready. I'm so ready," she said. "Every year, every Pap smear, it's been something. And these fibroids have caused havoc. I've had nonstop bleeding for weeks." Then a pause. "At the same time, though, there's a sense of loss."

"Well, that's understandable," I said. "It *is* part of your body. I mean, I get sentimental about losing my hair"—at which point Shannon snickered. "I practically weep whenever I look at the top of my head."

"Well, when you put it that way, I guess it's okay to feel sad about losing my uterus."

The inspiration for throwing the party had come, in small part, she admitted, as a reaction to something our mother had said. "Mom being Mom, when I first told her I was having the hysterectomy, she said right away, 'Now, Shannon, don't make a big deal of it.'"

We laughed. If the root of our family's dysfunction could be boiled down to one sentence, it would be: *Don't make a big deal of it.* How many times had our parents, now in their late seventies, given Shannon or me that admonishment? How many times had we done the opposite?

Shannon, according to family lore, entered the world creating drama. After Mom went into labor, my sister squirmed about and turned herself upside down, as if reluctant to leave the womb. An emergency Cesarean resulted, sparing mother and daughter a dangerous breech delivery. Henceforth Shannon was dubbed the child born backward, a characterization that endured and, unfortunately, sank in. Throughout childhood, she never felt good enough, smart enough, coordinated enough. Unlike Maggie, who possessed the grace of a natural athlete, or Colleen, who could be as poised as a beauty pageant contestant, Shannon was perpetually at odds with her body. This disharmony was never more apparent than when she had her period.

One episode burned into memory took place in the family car with me, Mom, and Shannon, who was thirteen. We'd been to the mall, though the purpose of the excursion and my reason for being there are forgotten. What remains is the tension in the station wagon as we drove home, the shopping trip scrapped because Shannon got hysterical. Mom had barely stepped foot into JCPenney when Shannon started sobbing and could barely walk because of cramps. My mother, who had precious little free time to shop, couldn't very well drag her bawling daughter through the mall or just leave her in the car, balled up in the fetal position. Mom's face as she gripped the wheel was a scramble of emotions—exasperation, concern, anger, and, I think, embarrassment. That Shannon's behavior occurred in public made it more egregious. I'll never forget how, back home, my mother, trying to be discreet, explained to Dad why shopping had been cut short: "She's in her way," Mom said, as if Shannon were a self-inflicted impediment.

To a degree, I think, Mom was dead on. Shannon did get in her own way. My parents took her several times to see Dr. Porter, who could find nothing wrong. The biological process that my mom and other sisters quietly managed remained a dramatic monthly struggle for Shannon. It was as though she had never moved beyond the frightful experience of her first period. My sympathy over the preceding year had settled into bewilderment. By this point in my boyhood, I well understood the notion of calluses. Why couldn't my sister toughen up?

If Shannon's tears didn't announce her time of the month, her wardrobe did. Immediately after getting home from school, she'd bag her body in the same oversized, pale yellow "granny dress," which she wore like a flag of defeat. In an odd coincidence, girls of the Spokane Indian tribe had historically worn their oldest dresses during menses, though the similarities probably ended there. Shannon retreated to her bed with a heating pad and bottle of aspirin, propping against pillows and taking up her embroidery. She was a Victorian spinster, prim and pitiful. I suspect her discomfort was symptomatic, too, of deeper anxieties about self-image and sexuality, which surfaced with the menstrual bloating and swollen breasts. She was a pretty girl, just over five feet tall, with flawless skin and a beautiful smile, but even on good days she carried herself as if boxed in by her body, hunched over, head down, arms strapping her bosom. I'm sure it didn't help that the older sisters teased Shannon about her plumpness, nicknaming her Circle. In a house with so many women, she felt isolated, and for the week of her period she withdrew from the family. I was scared for her but also a little scared *of* her.

Shannon's impending hysterectomy last fall revived family discussion of her history. "Why do you think it was always such an ordeal for Shannon?" I asked my older sister Maggie, whose own twelve-year-old daughter had just sailed through her first period, excited by this grown-up development. "She fought it," Maggie said simply. "She always fought it."

Shannon had a different answer. "It was that house. I internalized the tensions in the house," she told me. And added, as if in evidence, "They got better once I left for college." She also admitted to having been relentlessly naïve in adolescence. Though connected only by phone line, it was as if she and I were back in the yellow bathroom, her hand in mine. She repeated the same thoughts that had spun through her mind thirty years earlier: "It's supposed to be so natural, but you think, *Why is blood coming out of me every month? Something inside must be injured or wounded.* I mean, two tampons and a pad? Isn't it dangerous to lose all that blood?"

Another theory surfaced after her successful surgery. The surgeon informed Shannon that her uterus had been tilted at an unusual angle, and this pressure on her spinal nerves, added to a growing body and menstrual swelling, could explain what had exacerbated her monthly pain. "Now they tell me!" she exclaimed, laughing. "What timing! God, if only I'd known thirty years ago . . ."

I later wondered, would it have made a difference? Rather than consolation, the news might have simply added to her feelings of defectiveness, that yet another part of her was out of whack. But I'm glad not to have to recast the past for, buried beneath her childhood insecurities and extra weight, something truly remarkable was blooming: faith. Shannon, who had long been in open conflict with her physical body, began quietly and confidently embracing her spirituality. She was the only one of my sisters who attended Mass as frequently as Dad and me.

I had become an altar boy in third grade and served for five years. Every other month I was assigned to serve for a week at St. Augustine's 6 A.M. daily Mass, in addition to regular Sundays. Dad would drive Shannon and me to church, for he usually served as the lector. None of us ever got up early enough to eat breakfast and then fast the requisite hour before receiving Holy Communion, so we'd make the trip in a hungry, numbed silence. Morning after morning, we'd roll through the dark, empty neighborhoods and down the hill to church, as if in a recurring dream, one that I can still easily conjure.

I am twelve years old and walk three steps behind Dad as we enter St. Augustine's dim sacristy, dipping our fingers into the font of holy water. I hang up my coat and pull the blousy black-and-white cassock on over my clothes. Dad scans the Bible readings and speaks with Father Austen, who responds to his polite, whispered chitchat in a gravelly roar. I light the altar candles and pour water and wine into glass cruets, which I'll later have to carry—*don't spill, don't clink, don't trip, don't drop*—in the processional, a three-man parade (Father, Dad, and Boy) from the sacristy, into the foyer, down the side aisle to the front of the church, then up the main aisle to the altar. It is all quite showy considering that there are no more than two dozen of the devoted looking on—a huddle of nuns, a sprinkling of old people, and there, in the third pew, Shannon.

I am here for one reason: because Dad says so. Though not yet a disbeliever, I am skeptical. I've glimpsed too many of the goings-on behind the velvet curtain. Shannon, by contrast, attends Mass out of fascination. This difference in perspective is never clearer than during the hushed moment at the heart of the Mass when bread and wine become Christ's flesh and blood, the miracle of transubstantiation. From my vantage point, kneeling at Father Austen's feet on the right side of the altar, I can easily see Shannon's face. Her look is always the same, an expression of awe. When Father Austen holds aloft the Communion wafer in the consecration of the host, she is wholly enrapt in this retelling of the Last Supper, as if hearing it for the first time. I, on the other hand, can't help thinking of the cellophane bag in which a hundred hosts came packed, like potato chips. On autopilot, I ring the altar bell three times. Next, when Father Austen raises the chalice of amber wine, I see only the gallon jug from which it had been poured, stored under the sacristy sink, and can already smell the sourness that will later be on his breath. Again, the altar bell. In the quiet that follows, I watch Shannon, whose head is dipped, fingers pressed in prayer. She looks like she's captured a firefly in her hands and peeks to see its light.

The wine will not be offered to parishioners to drink, in part for simplicity's sake, I suppose, but also because Christ's blood already exists in the Eucharist, just as blood is present in human flesh. After taking a host himself, Father Austen places one on my tongue, where it is left to melt—never to be chewed—and I follow him to the altar rail. It's my job to hold both the basket of hosts and the long-handled golden paten under each recipient's chin, lest a host should fall. *But don't even think it!* A host must never touch the ground. Wafer catching involves keeping your eye on parishioners' tongues—gray-red, stubbled pelts, mostly, shooting out on waves of bad morning breath—a nauseating task on an empty stomach. To me, the sacrament of Communion means that Mass is almost over. To Shannon, second in line behind Pete the usher, it signifies far more. Her tongue slides under the chain-link fencing of her metal braces, and I momentarily meet her gaze.

"This is the body of Christ," pronounces Father Austen.

"Amen," she returns. In Shannon's flushed, avid face, I see gladness, as she is united with the Son of God.

I couldn't help but smile at her joy, though at the time I didn't fully understand it. She drew from a depth of feeling that I'd yet to form about anything. In this respect, she was far beyond me. The girl born backward had pulled ahead, and I couldn't have been happier for her.

In addition to attending daily and Sunday services, Shannon played her guitar and sang at Saturday's folk Mass. On a bookshelf in her room she kept a collection of nun dolls, each half a foot tall and dressed for a different religious order and historical period. I always found them creepy, with their Kewpie-doll faces and smothering garb, but to Shannon each represented a saint. For example, the doll in the white habit and black mantle, a Dominican nun, was Catherine of Siena. And the female Friar Tuck, dressed in the brown robes of a Carmelite sister, was Thérèse of Lisieux. Shannon also read *The Lives of the Saints* and developed a profound admiration for Teresa of Avila, the sixteenth-century Spanish nun and mystic. She took Saint Teresa's name when she was confirmed in eighth grade and also sought out copies of her books, which was unusual in itself. Unlike Ellen or me, Shannon wasn't much of a reader. In Teresa's life story, though, Shannon perhaps saw glimpses of herself. "I had one brother almost of my own age. It was he whom I most loved," Teresa wrote in her autobiography. She continued:

> We used to read the lives of the saints together; and,
> when I read of the martyrdoms suffered by saintly
> women for God's sake, I had a keen desire to die as

they had done. . . . I used to discuss with this brother of
mine how we could become martyrs. We agreed to go
off to the country of the Moors, so that they might be-
head us there. Even at so tender an age, I believe that
our Lord had given us sufficient courage for this, but
our greatest hindrance seemed to be that we had a fa-
ther and a mother. . . .

When I saw that it was impossible for me to go to
any place where they would put me to death for God's
sake, we decided to become hermits, and we used to
build hermitages, as well as we could, in an orchard
which we had at home. . . .

By rights, Shannon should've replaced me as the family
acolyte, but the Catholic Church forbade girls from service. This
ban was not reversed until 1983, and even then it was left to individual bishops to decide whether to integrate. The church provided no role for a girl like Shannon, except to sing in the choir.
And that was a new privilege in terms of church canon, only first
allowed in the early 1900s. In the preceding seven centuries,
with rare exceptions, no woman could wear a choir robe. She
could sing from the pews but, because the choir sang sacred
liturgical texts, only men were permitted. A legacy of Leviticus 15, this and many other anti-woman prohibitions officially
entered church law under the *Corpus Iuris Canonici* (1234 to
1916).

Pope after pope would reiterate that, because women bled
and were hence unclean and impure, they threatened the holi-
ness of the church. It goes without saying that, if they couldn't
sing in an official capacity, women couldn't become ordained,
distribute Communion, or serve as lectors. Nor could they touch
the chalice, the sacred vestments, or the altar linen upon which

the Eucharist was placed, even, I would suppose, to clean them. As for whether a girl or woman in her period could receive Communion, interpretation varied. In its strictest form, she would have to forfeit taking the sacrament, her abstention effectively announcing her menses to the entire congregation.

It was within this hostile environment that women such as Catherine of Siena (1347–1380) and Teresa of Avila (1515–1582) fought to create a role for themselves in the church. I can see now why Shannon was attracted to them. Beyond their saintly virtues, these were smart, articulate, confident women, who suffered greatly yet drew upon superhuman strength. In 1970, when Shannon was twelve, they became the first two women to be named Doctors of the Church, so honored for their extraordinary writings. In the same way that a teenage boy might live vicariously through comic-book characters, Shannon drew inspiration from these female saints.

Saint Catherine was bold. She spoke her mind. She led a life filled with adventure. Her impassioned voice can still be heard in her published letters and the celebrated mystical work *The Dialogue of Saint Catherine,* a transcript of a five-day rapture. What intrigues me most about this writing, and what I imagine comforted my sister, was Catherine's unflinching embrace of blood. It suffused her work. She saw blood as glorious, God's great gift to humanity through the sacrifice of His only son. But more so, she saw the souls of the faithful as blessedly drenched in it— bathing in it, even drowning in blood: "A man can possess the whole world and not be satisfied . . . until blood satisfies him." Her imagery could be sensuous and unabashed. In a vision Catherine later recounted to her confessor and biographer, Christ offered her the reward of drinking straight from the crucifixion wound at his side: She placed her lips "over the most holy wound, and long and eagerly and abundantly drank that indescribable and unfathomable liquid. Finally, at a sign from the Lord, she detached herself from the fountain, sated and yet at the same time still longing for more."

DESPITE THE CATHOLICISM IN OUR HOME, MY PARENTS GAVE EACH of us the option of attending public or private high school. As had older sister Ellen, Shannon chose Marycliff Academy, a small, all-girls Catholic school. When it closed in 1975 for lack of students, she transferred as a junior to Gonzaga Prep, the big coed Catholic school I was just entering. Though housed in the same building, the freshman and junior classes seemed to exist in separate counties. I was excited to finally be in high school, eager for weekend keggers, dances, football games. Most Friday nights Shannon could be found at home in her room, playing the guitar or doing needlework. I remember seeing her drift through Gonzaga's packed hallways. Her deep spirituality gave her an otherworldliness that made her seem woefully disconnected, like a girl suspended between heaven and earth.

A simple fact of human biology is that blood travels to the body's farthest extremes but always returns to the heart. So, too, with kin. Shannon's life and my life converged at the same spot in 1983, a pivotal year for both of us. I'd just moved to Seattle, having meandered through four years at Santa Clara University in California; she'd been living there since graduating from a small private college in Montana, where she had studied the one subject at which she'd always excelled, religion. We lived directly across the street from each other on Queen Anne Hill. Shannon and I saw each other often, sharing meals, going to movies, yet we could not possibly have been headed in more opposite directions. Unbeknownst to her or any other family member, I was coming out, dating men for the first time. At the same time, Shannon was following Saint Teresa's example, taking the first steps in becoming a novitiate of the Discalced Carmelite nuns, the cloistered order Teresa had founded in the mid-1500s. (*Discalced* means "barefoot," a defining aspect of their asceticism.) At age twenty-five Shannon was preparing to leave society, while, at twenty-three, I was finally emerging into it.

Our apartments reflected this divergence. Her studio was as Spartan as a monk's cell, merely a bed and a table with a single place setting. As is the prerogative of little brothers, I poked fun: "Jesus, Mary, and Joseph, have you already taken your vow of poverty?" My place, by contrast, was an overgrown hothouse, with grass-green shag carpeting, potted flowers, and walls covered with sister Maggie's enormous, bright-colored paintings. It was packed with thrift-store furniture, the air thick with Halston cologne. Madonna never left the turntable. I always made sure to hide the latest issue of *Christopher Street* when Shannon came over, visits that grew less frequent. Coming out was a nighttime vocation, I was learning, as I ventured to bars and clubs and occasionally ended up going home with someone. Shannon's spiritual life started at the crack of dawn, with sunrise service. What little else I knew about her training with the Carmelites didn't thrill me. *This is her choice,* I had to keep telling myself, *this is making her happy.* Still, one requirement seemed especially harsh: Once Shannon began living at the monastery, she'd have to sever all ties with family and friends for five years.

Although Shannon had known about other orders that ministered to the poor or worked in hospitals and schools, she could imagine no higher calling than devoting her life exclusively to praying for the betterment of the world, she has since explained to me. The only contact the sisters had with people outside the convent was through handwritten prayer petitions slipped into a narrow slot in the monastery wall.

If Shannon thought she'd be winning our father's blessing, she was mistaken. My parents both thought entering the monastery was a bad decision. She continued on nonetheless. Just as I was seeking my own community of acceptance, so was Shannon. She desired an authentic sisterhood, a connection with other women that she'd never had in our family. She also prayed that the cloistered life, free from the stress of the everyday world, would restore order and peace to her body. Blood, as ever, was a monthly ordeal. Though less painful, her menstrual cycle had become distressingly irregular. Shannon hoped the sisters' ascetic disciplines would be an anchor, physical as well as spiritual.

As the first stage in joining the order, she had to meet regularly, over a six-month period, with the monastery's Reverend Mother. The two women were separated by a small grille and, while it sometimes felt like being in a confessional booth, Shannon has said, she found the Reverend Mother to be a wise, wonderful woman with a delightful sense of humor. "We'd talk about my conviction in becoming a Carmelite, about prayer and faith. And about family. She was kind of like a therapist."

In Shannon's most recent retelling of this story, I found that it had more shades than I'd previously known. "Only one thing scared me about joining," she told me, trying to sound portentous but breaking down in laughter.

I tried to guess. "They wouldn't let you play Joni Mitchell on your guitar? No crewelwork allowed?"

"No, no, no. I was really nervous about—don't laugh—about my feet being cold. You had to wear sandals with bare feet. And," Shannon said, as if this were the final straw, "the monastery was unheated."

"So that was the deal-breaker?" I laughed. "The bad footwear?"

"Well, no—"

She fell silent. "Actually, you helped me with that," she said.

"I did?"

"Yes—" Another pause. "—when you came to me and shared that you were gay."

I wasn't following.

She then explained that she'd looked more deeply into the church's views on gay people, and that ended her desire to join the Carmelites. "In fact, that's when I left the church."

I didn't know what to say. I felt like I'd unwrapped an eighteen-year-old present, one made with such love, but also one that I could not have appreciated at twenty-three. I'd have seen it then as too huge a sacrifice, a debt I'd need to repay. Now I saw it not as an abandonment of religion but as her claiming her own voice, an expression of true faith.

"Well, there was no question," Shannon added. "I just turned completely around and walked the other way."

FIVE

Is he strong? Listen, bud!
He's got radioactive blood. . . .

—*SPIDER-MAN* THEME SONG LYRIC, 1968

MOST OF MY BEST INFORMATION ON THE ARCANE INNER
world of comic books has come from Steve's occasional late-night, half-medicated commentaries. "Listen to this," he said in bed recently, reading aloud a
snippet from the letters page of a back issue of *Fantastic Four:* "This is a fan writing: 'Dear Ladies and
Gentlemen, . . . The characters shine. They live and
breathe. Real, red human blood pumps in their veins.
I can easily believe their world for, though colorful
and bizarre, it is just as real as ours. . . .'"

Steve chuckled. "Wow, he's got it bad."

I put down my *New Yorker.* "Fantastic Four? That's the one with Mary Hart?"

"Uh-huh."

Of course, this made perfect sense to me. A young Mary Hart (from TV's *Entertainment Tonight*) would be Steve's pick to play the Invisible Woman in a big-budget Fantastic Four movie.

Steve then commented on the fan's use of the word *real* to describe a comic with a woman who can turn invisible, her brother who can burst into controlled flame, her husband whose body becomes Silly Putty, and her friend who's essentially an animate pile of rocks. I still wasn't sure, though, whether Steve was poking fun at the letter writer or being serious.

"Cool, isn't it?" he said. Okay, there was my answer.

"Yeah." It *is* cool. I find the wholehearted suspension of disbelief that avid comics fans share to be marvelous. Although I prefer the sure footing of nonfiction, I still envy that fearless willingness to lunge into pure imagination. My beloved collections of Joan Didion essays are never so transporting. The degree to which a comic-book reader is drawn into the illusion depends upon the adherence to a set of conventions dating back to the late 1930s, the earliest days of this indigenous American art form. The heroes must have fabulous powers or abilities. They have bright costumes and dual identities. The conflict between good and evil is clearly delineated. And the convention that wraps all these elements into one neat package is the origin story, the tale of a character's pivotal moment of transformation. Whereas ancient myths always have a definite resolution—odyssey's end, deification, betrothal, and so forth—superhero comics are usually meant to be never-ending sagas. Regardless of the adventures yet to come, though, the character is always anchored by his or her origin. Superman—no matter what a current creative team does with him—will always be a survivor of the doomed planet Krypton, raised by the Kents in Smallville, Kansas.

In the life of a comic book, the origin story may be retold dozens of times. Usually it's done in a flashback, deftly recapped in a handful of panels, often to jump-start a new storyline. Marvel Comics, mindful that each issue might be a reader's first, used to summarize the title character's origin in a box on the splash page. I love these old thumbnail bios. Allow me to introduce, for instance, the original Spider-Woman:

> *When Jessica Drew's father injected her with a serum of spider blood, he cured her of a fatal disease ... and changed her life completely! Watch her, now, as she confronts her responsibilities, problems and unbelievable POWERS!*

The idea that heroes often have supercharged blood reflects the real-world belief that qualities course through our blood. There being no scientific evidence for this does not dispel the notion. When the anchor of the nightly news praises firefighters for the "bravery pumping through their veins," we don't disagree. Heroes, whether actual or fictional, seem to have a blood type the rest of us don't. This conception is amplified manyfold in comic books. Captain America, for instance, has Super-Soldier Blood; his courage isn't the courage of one man but that of an entire battalion. Such is the potency of superhero blood that a transfusion from the original Human Torch (who, by the way, wasn't even human) helped transform the character named Spitfire into a superspeedstress. Then there's the She-Hulk, formerly a petite attorney. Near death after a catastrophic car accident, she was saved by her cousin, Bruce (aka the Hulk), who gave her an emergency blood transfusion, unintentionally turning her into a female version of himself, the Savage She-Hulk.

Creative teams do sometimes bend the rules. Some superheroes wear street clothes rather than costumes, for instance. Likewise, some have powers that do not originate in their blood but come from an external source, such as a talisman or exoskeleton. And there are others who don't even have origin stories. Or, to be more precise, whose origins are shrouded in mystery. Of these, the superstrong Savage Dragon comes to mind. Dragon has a healing factor that enables him to recuperate from any injury, yet he doesn't know how he acquired it. His earliest memory is of awakening in a burning field, naked, a full-grown man, bright green, with a large fin on his head. Despite this blank slate, he was driven to do good, as if heroism were encoded in his DNA. Whether he realized it or not, destiny clearly had plans for him. It is this particular narrative—the rise of an unknowing or unlikely hero—that I'm drawn to in my reading, the true stories of individuals who simply followed their passion and somehow ended up making history. Such a man was Antoni van Leeuwenhoek.

AT THE SAME TIME THE DUTCH ARTIST JAN VERMEER WAS PUTTING finishing touches on his last great painting, *Allegory of Faith*, his lifelong friend, the naturalist Antoni van Leeuwenhoek (1632–1723), was in a nearby studio quietly discovering a new universe, one of previously unimagined marvels—that of microscopic life. Using a small microscope of his own design, he was the first person to observe, draw, and describe what he called "very little animals" (now known as microorganisms), including the bacteria swimming in human saliva, the protozoans in pond water, and the sperm cells in semen. Likewise, he discovered red

Antoni van Leeuwenhoek

blood cells, an accomplishment that changed the way scientists regarded the blood, transforming it from a simple fluid imbued with unseen spirits and qualities to one of burgeoning complexity. In addition, Leeuwenhoek (commonly pronounced *LAY-when-hook*) contributed to the understanding of capillaries, the newly discovered vessels bridging arteries and veins, and documented similarly intricate structures in the roots, stems, and leaves of plants. He is revered today as a father of multiple disciplines: microscopy, microbiology, botany, and hematology.

To fully appreciate these achievements, however, one needs to know about Leeuwenhoek's humble beginnings. Imagine, if you will, someone like the owner of your neighborhood dry cleaner, the polite but taciturn man whose tidy little shop you patronize now and then. He is a stocky fellow with blunt features—bulging, heavy-lidded eyes, a bulbous nose. You've heard that he's a widower who has also lost several children to illness, which may account for his sad air. His one surviving child, a daughter, helps him in the shop, which does modest business. To make ends meet, however, he must do janitorial work on the side. You've scarcely ever seen him out strolling the neighborhood. He and his daughter live in the flat right above the shop, where, word has it, he spends every moment of his spare time tinkering, always tinkering. Late at night you may've glimpsed his silhouette against the upstairs drapes. This is a snapshot from the early 1670s of the life of Leeuwenhoek: a curious, hardworking man, an accidental scientist.

Born in Delft in 1632, just a week before Vermeer, Antoni lost his father, a basket maker, at five years old, and his mother at age eleven. At sixteen he moved to Amsterdam to apprentice in the cloth trade. With scant education to speak of and knowing no language but his native Dutch, he did have an ability that served him well—a gift for mathematics. He returned to Delft and opened a fabric shop in 1654, the same year he married his first wife, Barbara. The next dozen years took a heavy emotional toll on Antoni. Only one of his five children survived past age two—his daughter Maria—and Barbara died in 1666. Shortly thereafter, he began experimenting with microscopes, out of curiosity's sake, to be sure, and perhaps also, it occurs to me, as a way to fill the lonely hours of the night. Keeping busy may have also helped. He had side jobs as a land surveyor and wine assayer, and he continued at his long-term post as chamberlain (a glorified janitor) for an office of local sheriffs.

In hindsight, it appears that everything Leeuwenhoek lacked—formal education, professional ties, personal fortune—worked to his benefit as a scientist. When he wished to look through a microscope, he had to construct his own since he couldn't afford one. He even learned how to blow glass and, in the process, became a master at grinding lenses. Lacking a sophisticated vocabulary and being "quite a stranger to letters," as one colleague later wrote, Leeuwenhoek had to invent terms to describe his uncanny observations. Hence, his "little animals," which efficiently conveyed that these bacteria, protozoans, and spermatozoa were indeed living creatures. Though he had no talent as a draftsman, he made do at first with his own crude sketches, which honed a skill for memorizing visual detail, useful in comparing countless specimens. Unfettered by preconceived notions, beholden to no one, he was poised to break new ground, ". . . for being ignorant of all other Mens thoughts," wrote Dr. Thomas Molyneux in 1685, "he is wholly trusting to his own."

According to one of my favorite scientific tall tales, the first microscope was "invented" in the mid-1500s by some unnamed lunkhead who mistakenly used his telescope backward. "Land ho!" became—*bump!*—"Oh, land," and a new instrument was christened. But the true story is a lot more complicated. Its origins can be traced as far back as the earliest recorded descriptions of optical phenomena. Magnification by curved transparent surfaces was recognized by the first-century Roman philosopher Seneca, for instance, who wrote that "letters, however minute and obscure, are seen larger and clearer through a glass bulb full of water." This effect was also created by polished gems, as reported around the same time by Pliny the Elder, who noted that the nearsighted Emperor Nero used an emerald to improve his vision while watching gladiatorial contests. The height of decadence, it seems to me, Nero's emerald monocle must've been both effective and stylish, but there's no evidence that he launched a trend. Which stands to reason. Not to be myopically insensitive, but one cannot envy what one cannot see. It would be another twelve centuries before the use of concave lenses for the deliberate purpose of enhancing eyesight was proposed, credit for which goes to the English monk Roger Bacon, who in his encyclopedic *Opus* of 1267 also predicted the invention of the microscope. His contribution to the field of microscopy was only acknowledged in retrospect, however. The monk was imprisoned for heresy, and his writings remained undiscovered until the eighteenth century.

The invention of spectacles as we know them today was

made independently around the year 1285 in Florence by a man named Salvino degli Armati, a fact that, oddly, wasn't made public until after his death some thirty years later. It seems that, like a well-guarded family recipe, he shared his creation with only a select group of friends. Subsequently, though, the use of eyeglass lenses took hold and spread throughout Europe. And it was only a matter of time before someone, rather than placing lenses side by side, arranged them one before the other, thus creating a compound magnifying instrument. Official recognition for the first microscope, however, is often ceded to a Dutch spectacle maker, Zacharias Jansen, who in 1590 combined two curved glass lenses in a small tube as a means for studying minute objects. Seventy-five years later an Englishman, Robert Hooke, stirred the public's imagination with his startling book on microscopy, *Micrographia* (1665). In it Hooke described and illustrated what he'd observed using his own compound microscope—the hairs on fleas and snow crystals, for example. He also unknowingly coined a new scientific term when writing on why cork floats. Under magnification, the tiny air pockets he saw looked like the small rooms in monasteries, commonly called cells. Hooke had no idea at the time that he'd discovered plant cells.

One noteworthy person who picked up a copy of *Micrographia* was Antoni van Leeuwenhoek. Though it's doubtful the Dutchman could have read the English text, the lush engravings of the small-made-big must have made his brain itch. He began to tinker. Rather than duplicating the elegant but complicated two-foot-tall microscope Hooke had drawn in the book, he went in the opposite direction. Borrowing the basic design of the magnifying glass he used in his shop to inspect the weave of fabrics, he crafted a lightweight, handheld device that housed a single lens. What to others may've seemed like a step backward was actually a great advance. The microscope lenses in common use at the time were made of poor-quality molten glass and had a power to magnify an item just twenty or thirty times. When viewed, objects appeared to be surrounded by fringes of color, and with each additional lens the optical defects multiplied. By using one spherical lens, ground and polished from a bead of purer glass, Leeuwenhoek found he got far clearer images and a magnification of more than two hundred times.

When I recently held a replica of one of Leeuwenhoek's microscopes, my first thought was, *It sure ain't pretty.* It was punier than I'd expected, and the lens was—no pun intended—microscopic. But I held my tongue. I did not want to offend the man who'd made it, Al Shinn, who sat across from me in his ramshackle cottage in Berkeley, California. Al, in truth, had done a magnificent job in re-creating this seventeenth-century microscope, I knew, basing it on a Leeuwenhoek original preserved in the Utrecht Museum in the Netherlands. A specialist in designing high-tech optical instruments, Al had spent years studying the Dutchman's notes and designs, experimenting with different metals, even replicating the device's tiny screws, scoring the threads by hand.

For all the excitement it generated in its time, the design is pretty simple. A tiny lens—a two-millimeter glass bead—is held between two thin brass plates, which are riveted together. By way of a long screw, you hold the device up to your eye like a rectangular lollipop. The object you want to view is affixed to a metal pin on the back that can be rotated or repositioned using a second screw. Leeuwenhoek produced more than five hundred variations on this design during his lifetime and bequeathed the bulk of them to his devoted daughter, Maria, who had never married and who'd assisted him till his dying day. Following her death in 1745, they were all sold at auction, as per Maria's request, yet only nine are known to exist today. This is sad but not surprising. An untrained eye would never guess their purpose. The replica resting in my palm looked like an obsolete carpentry tool, something you wouldn't hesitate to toss from a junk drawer.

When Al had greeted me at his front door—my first time meeting him in person—I'd immediately thought, *It's Doc,* the scientist played by Christopher Lloyd in the *Back to the Future* movies, the inventor of the time-traveling DeLorean. Al, who's sixty and gray, had the same detonated hair, lively eyes, and endearing smile, as well as a similar lankiness. Wearing a T-shirt, sweats, and flip-flops, he cleared a path back to the kitchen, where he made me a nice strong cup of coffee. Al's house, like the physical manifestation of an active mind, was filled with stuff: projects that appeared to be just started, half finished, or long since abandoned. I even spotted boxed chemistry and ham-radio kits that could've been from his 1950s boyhood. While he shares the two-bedroom home with his wife and only child, a teenage daughter, I saw scarce evidence of their belongings. It was easy to imagine his many, many interests squeezing out any of their own. In the press for space between old computers and piles of books and laundry, a lampshade doubled as a bulletin board, covered with Post-it notes. I pushed aside some newspapers on the couch and sank into a cushion.

As if to explain away the surroundings, Al admitted, "I've always been a tinkerer. Even since I was little, I've been interested in the instruments of science—radios, telescopes, microscopes." He'd first been inspired to try replicating Leeuwenhoek's microscope about ten years ago, he continued, while he was working for a Bay Area ophthalmic equipment company, Humphrey Instruments. At the time, Al was a principal research scientist for the firm, a high-level position he'd reached more from raw skill than from schooling. A college dropout in the early 1960s, he had made his living for many years as a "hippie jeweler—pipes, earrings, artsy-craftsy stuff," a line of work that had drawn him from Maryland to, go figure, the Berkeley area. After Al married, his new brother-in-law informed him that Humphrey was hiring. He got a job on the assembly line and rapidly worked his way up, the ascent fueled by a natural facility for designing and building complex optics systems.

When Al left the company a few years later to pursue freelance work, he found that his high-tech experience had in no way diminished his love of the low tech. He finally had a chunk of time to perfect his replica and to engage in a little hero worship. "Everyone else in his day was looking at small things and making them bigger," Al explained, "but Leeuwenhoek was the first guy to look for the invisible—what's there that can't be seen. He started with things like pond scum."

"And blood, right?" I interjected. "Weren't red blood cells among his earliest discoveries?"

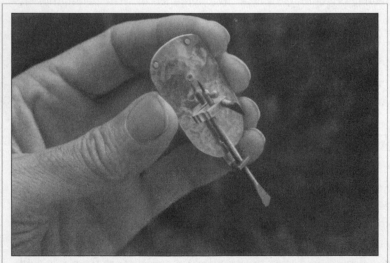

Al Shinn's replica of a Leeuwenhoek microscope

That's right, Al said.

"Well, um, do you think it would be possible for me to do that—to look at some of my blood through your microscope?" I was suddenly embarrassed that I was proposing myself as a specimen. "You know," I added, "to see my"—I now thought it best to use the scientific term—"corpuscles? Would it work?"

His face lit up. "I've never tried it before. But there's only one way to find out," he declared, "and that's to take the experimental approach!"

"I brought a needle," I offered helpfully.

"Really? You brought a needle?"

"Yeah, a sewing needle. To prick my finger?"

For a brief second, nothing, then a smile swept across his face. "Excellent!

"Okay," he added, "let's see now, somewhere here I have some microscope cover slips . . ." And Al was off, as if he heard the pinging of a tracking device somewhere in the distant clutter. I, meanwhile, gave the Leeuwenhoekian lens a good long look.

In the year 1668, around the time he started experimenting with microscopes, Antoni van Leeuwenhoek began attending public meetings held weekly in Delft by a group of area doctors. Here he witnessed autopsies, heard lectures on new areas of scientific and medical investigation, and eventually submitted for consideration his own fledgling findings. His reports caught the attention of a participating doctor, Reinier de Graaf, who was also a member of the Royal Society of London, an association of progressive European scientists, including such illustrious figures as Sir Isaac Newton. The Royal Society had the standing practice of urging its members to write with news of their important discoveries. In 1673 Dr. de Graaf happily obliged, his great discovery being not an idea or a technique or an innovation, but a man, ". . . a certain most ingenious person here, named Leeuwenhoek."

With introductions made, Antoni was thereafter invited to correspond directly with the society, a practice he would continue for fifty years. His many letters, written in colloquial Dutch, were translated into English and published in the Royal Society's prestigious journal. Although his reports were often rambling, there was no mistaking the originality of the amateur's research. What came through just as clearly was how fearless Leeuwenhoek was in attempting to see what no one had seen before. And I do mean fearless. He had wished, for instance, to watch gunpowder explode under his microscope, so he devised a contraption for viewing the fireworks up close and, although he nearly blinded himself, succeeded. In another wild experiment, Leeuwenhoek determined to answer the question, *Why is pepper so hot?* He mashed peppercorns, soaked them in melted snow (thought to be 100 percent pure water), and, several days later, prepared a sample for his lens. As he wrote in the spring of 1676, he fully expected to discover in the magnified pepper particles "sharp little needles," which literally lacerated the tongue. Instead, Leeuwenhoek found something wholly unrelated—four different kinds of "little animals" swimming in the sampling. The first three were protozoans, the organisms he'd previously seen in pond water, but the fourth set of creatures darting about was new, a separate and much smaller breed. They resembled tiny eels, he observed, "lying huddled together and wriggling" or "moving about in swarms." Leeuwenhoek, we now know, had discovered bacteria. (The pepper's heat, by the way, went undiagnosed.)

Leeuwenhoek found the same "little eels" in human saliva and other substances, he reported in later correspondence to the Royal Society. Soon scientists, clergymen, and common folk were making the trip to Delft to see for themselves the cloth merchant's menagerie. That we are surrounded, covered, and filled with countless creepy-crawly microorganisms is such a commonplace understanding today that it's difficult to imagine how radical—and grotesque—Leeuwenhoek's images must have seemed three-hundred-odd years ago. His letters are a bracing reminder. "I have had several gentlewomen in my house, who were keen on seeing the little eels in vinegar," he wrote in 1683, "but some of 'em were so disgusted at the spectacle, that they vowed they'd ne'er use vinegar again. But what if one should tell such people in [the] future that there are more animals living in the scum on the teeth in a man's mouth, then there are men in a whole kingdom?"

Although Leeuwenhoek always strived to be hospitable, the flood of visitors intruded on his precious work time. During a single four-day period, he once bemoaned, he received twenty-six separate callers. But not all drop-ins were unwelcome. Leeuwenhoek's makeshift laboratory became a mandatory stop for visiting royalty and heads of state, including King Frederick I of Prussia. Queen Mary of England arrived unannounced one afternoon, but the Delft shopkeeper was not at home. Leeuwenhoek was crushed. This missed meeting, he wrote, "will, and must, be mourned by me all my life." Henceforth, appointments became mandatory. Another memorable visit came from Russia's Czar Peter the Great, for whom Leeuwenhoek demonstrated all manner of "microscopical observations," including, as a grand finale, the movement of blood through the newly discovered capillaries. This never failed to astound people. Leeuwenhoek had designed a special microscope to which he could fasten a small, live fish. Because some fish have transparent tails, one could see blood traveling through the microscopic "tubes" connecting the tiny arteries to the tiny veins. The Russian monarch, who spoke some Dutch, was "so delighted," a local historian wrote at the time, "that in these and other contemplations he spent no less than two hours, and on taking his leave shook Leeuwenhoek by the hand, and assured him of his special gratitude for letting him see such extreme small objects." Leeuwenhoek returned the compliment by presenting one of his microscopes as a gift, something he rarely did. Neither did he ever sell his microscopes or teach others how to make them. Anyone who wished to see a Leeuwenhoek microscope had to pay the man a visit.

"Here we go!" Al called out to me from the hallway. With a small box in hand, he returned to his seat and began fashioning a tiny slide from two thin pieces of clear plastic. "Okay, what we're gonna do first is mount these on the pin with a little bit of beeswax. Oh, look at that!" he said, delighted, halting his work. "Newton's rings."

"Huh?"

Al held the plastic shards before me. "Do you see the rainbow-colored rings there, like oil in water?"

"Oh, yes," I responded.

He smiled. "That's caused by the interference of the light between the two surfaces. One of Newton's discoveries."

Even as he struggled to mount the plastic slide, Al remarked, "We have it so much easier than Leeuwenhoek did. What did he have to do to get a thin piece of flat glass? Window glass would be way too fat, so he had to make it himself!" The trouble Al was having made me realize why Leeuwenhoek got into the habit of leaving difficult-to-mount specimens permanently in place, then making a fresh microscope.

At last Al succeeded in getting the miniature slide onto the Leeuwenhoek replica, a process that had taken a good twenty minutes. And even then, Al wasn't sure it would work. If the slide was too thick or not balanced just so, the specimen would be too far from the tiny lens and impossible to draw into focus. "I don't want to waste your finger prick," he said. "Let's try some scummy water first." He did not even have to leave his chair to find some—a vase on the coffee table held flowers that had been dead at least a month. "That looks good and slimy," he said, raising a stem to retrieve a drop.

Its looks were deceiving, however, as Al could find no "wee beasties" swimming in the sample. Drop after drop of smelly water he tried but, alas, nothing surfaced. *Yuck,* I remember thinking. *Is there really water so gross that even bacteria will refuse to move in?* He did not look ready to surrender, so I spoke up. "Al," I said. "Let's go for blood."

While he cleaned the slide with his shirtfront, I pulled out my pack of sewing needles, selected a medium-sized one, and poked my index finger. Though I'd pushed hard, no blood appeared. I tried another spot farther down. This triggered a fleeting flashback to high school biology, though it was my lab partner's blood we studied, not mine—a lucky flip of the coin.

"Ah, there you go," Al said, nodding, for he'd noticed before I did that both pricks were now bubbles of blood. "More than we'll need."

"Okay," he coached, "now get some of it on the end of the needle. Not much, just a bit." I handed the dipped needle to Al, who smeared the blood on the slide and gingerly lifted the microscope to his eye.

In one of his very first letters to the Royal Society, dated April 7, 1674, Leeuwenhoek noted, "I cannot neglect this opportunity to tell you that I have endeavored to see and know, what parts the blood consists of; and at length I have observed, taking some blood out of my own hand, that it consists of small round globules driven through a crystalline humidity of water."

(By "water," he was referring to what is now called plasma, the pale liquid in which blood cells are suspended.) Writing again two months later, Leeuwenhoek elaborated, not only describing "the red globules of the blood" but also measuring them. This was standard practice for Leeuwenhoek—he fastidiously measured everything he studied—and another of the man's innovations, making him the founder of the science of micrometry. The impulse to measure seems to have been perfectly natural to him, a numbers whiz, as well as an extension of his years as a cloth merchant and surveyor. In order to measure particles on such a small scale, Leeuwenhoek had to devise new means of comparison, such as using a single hair or a grain of sand. Thus, he reckoned that one red cell was twenty-five thousand times smaller in volume than a fine grain of sand—or about $\frac{1}{3,200}$th inch in diameter. Modern measurements indicate that he was almost dead on.

Leeuwenhoek returned over and over to studying blood, constantly refining his understanding. He compared human blood to that of different animal species and noted, correctly, that the "globules" are the same size in all red-blooded creatures regardless of total blood volume, whether in a minnow's thimbleful, say, or a whale's many gallons. In addition to discovering red cells, Leeuwenhoek described the coagulant properties of blood and made preliminary observations of the colorless corpuscles now known as white cells, or leukocytes. None of which is to suggest that Leeuwenhoek didn't make errors. Sometimes when you go out on a limb, the branch snaps. For instance, for the remainder of his life, Leeuwenhoek insisted that red blood cells were spherical when, in fact, they're more like pinched balls of Play-Doh or, if you will, like sunken jelly doughnuts. Further, in attempting to explain where in the body blood cells originate, he theorized that they formed from minute particles of food that were built up then rounded in the steady rush of the bloodstream, like pebbles polished by the tumble of waves.

He never hesitated to share the smallest details of his findings, and at times, in his nearly four hundred letters to the Royal Society, Leeuwenhoek was forthcoming almost to a fault. "Whenever I found out anything remarkable," he explained in 1716, "I have thought it my duty to put down my discovery on paper, so that all ingenious people might be informed thereof." A noble credo, but he did make exceptions. He never divulged exactly how he ground and polished his amazing lenses, for instance, or, as important, how he so successfully illuminated his specimens. He even admitted to having a couple of microscopes with lenses of superior magnification, a private cache he showed no one. This last bit of withholding, I believe, is a forgivable indulgence. I'd like to think that, at least for a short while, before the next generation of microscopists brought forth their own innovations, Antoni van Leeuwenhoek had the best eyes in the world.

"Whaddya see?" I asked Al. "See any red?"

Silence for a minute as he fiddled with the focus, and then, with a sense of ceremony, he handed his microscope to me.

I was surprised by what I saw, which pleased me most. My first impression was not of color, as I'd expected, but of translucent shapes: countless clear granules where I thought I'd see brilliant scarlet beads. They had a slushy appearance, as if I were looking at ice-frosted glass. At the edges, however, where cells were piled up, there was an unmistakable rosy tinge.

I was completely satisfied with the demonstration. Al wasn't. "Can you spare another drop?" he asked.

Al then disappeared with a new slide, freshly smeared with a touch of my blood. A few minutes later, he hollered from his back porch: "Come take a look! We've got a better microscope!"

Al had dug up a modern-day compound microscope, which he'd positioned atop his washing machine in the sun-drenched room. "This light is so perfect," he exclaimed.

With magnification set at about five hundred times—more than twice that of his Leeuwenhoek model—I could now see hundreds of my red blood cells, sharp and delineated. Most were stuck together, huddled as if for protection from my huge peering eye. But a few lay flat, perfect specimens. And if I wasn't mistaken, I also spied a couple of cells near the top of the slide with a different profile—white blood cells.

"Pretty cool, huh?" Al said.

"Very cool."

"It's just too bad that there wasn't anything crawling around in that water drop. But we can try that again, too—have a better look with this 'scope." Al surveyed his overgrown property through the back screen door. "There's got to be some scummy old water back here someplace," he said. "Let's see now . . ." With that, he set forth, blazing a trail through the wilds of his Berkeley backyard, in search of his own very little animals.

THAT OUR BLOOD ABSORBS FRESH AIR WITHIN THE LUNGS AND THEN circulates it throughout the body was first proved during Antoni van Leeuwenhoek's lifetime. Exactly how the blood carries out this task of transporting and discharging oxygen remained a mystery, however, for another two centuries. In studies conducted in the mid-1860s, a German pathologist discovered that the main component of the red blood cell is a complex protein he named hemoglobin, which gives blood its characteristic color. If you stop and think about it, the concept is counterintuitive: Blood is bright red *because* it is fully oxygenated, yet oxygen is, by definition, colorless. But this scientist proved that hemoglobin is actually a functioning pigment, the precise shade of which is determined by how much breath, so to speak, the cell is holding. To picture how it works, think of a balloon not yet inflated. It's deep burgundy in color. Now blow. As it fills with air, its color is stretched and the burgundy brightens to cherry red. This is hemoglobin in action.

Over the past hundred years a series of scientists have unveiled further clues as to the workings of the red cell (also known as an erythrocyte, *i-RITH-row-site,* from the Greek for "red," *erythrós*). With its supple disk shape, the erythrocyte can dock next to other cells in tissue throughout the body to perform the equivalent of mouth-to-mouth resuscitation. It not only breathes oxygen into these cells but also sucks up carbon dioxide and carries it back to the lungs. Its compact shape, together with its elasticity, allows the erythrocyte to squeeze through the narrowest of capillaries and then spring back to normal size. (When red cells have the wrong shape, as in people who have the hereditary disease sickle-cell anemia, the elongated and curved—sickle-like—cells cannot pass through capillaries; the resulting blockages cause intense pain and serious deficiencies of oxygen to tissue.)

A healthy erythrocyte will continue its nonstop travel, cycle after cycle after cycle, for about 120 days until, too exhausted and battered to go on, it drops off the circuit. Scavenger cells in the spleen gobble it up and strip it of iron and other components, which are then sent for recycling to the body's blood incubator: the bone marrow. Here, in these hot, spongy, fat- and vascular-rich tunnels, reside the ancestral cells from which erythrocytes, as well as all other cells of the body, are derived: the stem cells. Those that are specially programmed to become red cells divide and multiply; three million form per second. But these proto-red-cells, technically called erythroblasts, are not yet ready to enter the bloodstream. They must first mature and ac-quire ample hemoglobin. When they are finally ready to squeeze through tiny blood vessels and enter into circulation, a defining event occurs: They lose their nuclei, the cellular "brains" in which DNA is encased. Structurally, this sets erythrocytes apart from most other cells. Without a nucleus, there's more room for hemo-globin and, therefore, for more oxygen. But its absence also guar-antees each individual red cell's demise, for without a nucleus it cannot reproduce. What's more, stripped of DNA, the body's bio-logical signature, it has no identity. In this regard the formation of an erythrocyte is the antithesis of a classic origin story.

In the comic-book definition, the origin encapsulates a char-acter's pivotal moment of transformation, telling in orderly pan-els and with pithy phrasing how he or she came to be. Ordinary human beings have such tales, too. But we don't call them origin stories, we call them ordeals—those life-changing episodes that, assembled in hindsight, tell us who we are. Coming out in my early twenties was mine. For my sister Shannon, leaving both the church and her training with the Carmelite nuns back in 1984 was but a prelude to hers. The events that would truly transform her life began unfolding a few years later.

She was twenty-nine, living by herself in Seattle, and working as a department-store shoe salesperson. I'd moved to San Fran-cisco three years earlier and was living in the Castro district. In many ways Shannon and I were on similar paths—both lapsed Catholics, both dodging our disapproving parents at every turn, both trying to figure out who we were at the core—but going in different directions and at very different speeds. I had chased down a life I'd desired for years, whereas she, partly out of loy-alty to me, had relinquished hers. While I'd found my commu-nity, Shannon was now without one. The farther she'd gotten from Catholicism, the greater her disenchantment with its dogma. On the one hand, distance had made things clearer—she'd lost her religion but not her faith—and yet, stripped of her long-held identity and without a new passion, she struck me as being a little aimless.

During our regular phone calls, I always dominated the conversation with stories of my friends and my jobs—first at a small theater company, then at the modern art museum—and of my first serious relationship. That dynamic changed with a phone call in the spring of 1988. "I'm pregnant," Shannon told me.

I hugged the refrigerator. What came to my lips was not *Congratulations!* or *How wonderful!* but: "Are you sure?"

She was not only sure, she was more than six months along.

It hit me that I knew next to nothing about my sister's social life—aside from the girlfriend with whom she went to movies and out dancing—let alone her love life. "So, who is he?" I ventured.

A man she'd met at a club, Shannon answered. Someone she'd really connected with, at first. He was beautiful and Cuban and mysterious. And secretive and insistent and controlling. And scary and—more than anything else—dead set on getting a green card. He'd disappeared, who knows where.

My heart sank. "I am so sorry, Shannon, so sorry. Is there something I can do?" I was offering both consolation and an apology. I felt terrible that before now she hadn't been able to share all this with me, that I probably had not ever given her an opening. Regardless, I was the first in our family to hear that she was expecting. She'd just gotten the news herself earlier that week, Shannon said. Okay, now I was just plain confused. Didn't she say six months?

"I thought I was off again," she started to explain. "You know I've never had regular periods. It wouldn't have been the first time I'd gone a while without having one." This sounded like a bad rough draft of an answer, one she wasn't convinced of herself.

As for her swelling belly, she added, fumbling, "I thought I was gaining weight again. I thought I was having more stomach problems. I—" She stopped herself. "The truth is, I couldn't face the reality. I couldn't go through that door."

I was glad we were separated by hundreds of miles of phone line because the look of horror on my face would've killed her. Shannon had always been at odds with her physical self, but this was on a whole different level. With a calm I did not feel, I quietly asked, "Are you dealing with it now? You've seen a doctor, right?"

"Yes, I'm all checked out. I had a sonogram, *amnio*-whatever, everything. They say the baby seems to be fine."

"That is great. And you? How are you doing?"

"I'm doing . . . okay."

"Good, good." I sighed relief. "So, you're . . . You're keeping the baby, right?"

Shannon stated evenly, "I'm having the baby, but I'm not keeping it. I'm going to give it up for adoption."

Evidently, once her denial had been punctured, absolute clarity had kicked in. Shannon had already made arrangements for the adoption through a private agency recommended by her doctor. *Are you really sure about this?* I was tempted to ask again, but I kept quiet. Her voice was firm; she was not calling for advice, I could tell, but to let me know how everything was settled. The prospective mother, who had just written Shannon a warm, earnest letter introducing herself, was a family-practice physician, a single Caucasian woman who spoke fluent Spanish. Shannon chose not to meet her in person or to maintain any contact after the birth—feeling this would make it easier to separate from her baby—but she knew what she most needed to know, that this woman would provide a loving home. Plus, Shannon's child would have an older brother, for the mother already had an adopted little boy of mixed race. This woman had also offered to defray the costs of my sister's medical care, counseling, and other expenses, which lifted a burden since Shannon made little money and had no savings.

Despite the spinning in my head, I found myself saying reassuring things: She'd made the right decision; it sounded like a perfect match. But I held back what I most felt: Was there really no other way?

After speaking with me and my other sisters, Shannon broke the news to our parents. If she'd written a list in advance of all the things *not* to demand of her, it would have consisted of my father's three commands: She must return to Spokane and stay with them. The adoption must go through a Catholic agency. And the baby must be baptized.

Sorry, but no, no, and no. In so answering, Shannon was cutting her ties with Mom and Dad. My sister had never before exhibited such grit. White-knuckled at the wheel, she saw her destination eighty days ahead and wasn't allowing any obstacles to get in her way. Distancing herself from family members was crucial. At the time, two of our older sisters each had infant sons, the family's first grandkids; for Shannon, seeing them was too painful. She also insisted I not come up to Seattle.

"I went into hiding," she admitted to me recently, the first time we'd sat down together and spoken of this at length. She gave a sigh and a look that said, *I cannot believe how much I've changed.* "I did not want to see or be seen by anyone, including you." Among all her siblings, Shannon told me, my opinion had always mattered most, and she couldn't bear to have me witness her shame. But I think she was also protecting herself and her baby, in an almost primal way. It was as if she had physically grabbed the walls around her and pulled them closer in, forming a space she could manage, a place in which her tiny flame of strength could glow.

Two weeks after she'd first called with her news, my good friend Peter died. This was not unexpected. This wonderful, witty man—a diminutive and Deutsch Oscar Wilde, if you can imagine—had been sick with AIDS and bedridden for weeks at his home, where I and a corps of friends helped nurse him. Even so, I was stunned. This was my first experience of profound loss—something my sister was preparing for, too, I realized in the following days. Shannon, who'd met Peter during a visit to San Francisco, sent a note after she'd heard of his passing. Her words conveyed loads of sisterly love. "As it came time for him to let go of his life," she wrote, "it is my honor to give birth to a new one," a sentiment I found genuinely comforting. That she had pulled herself up to a place of such pride buoyed me at a moment when I was sinking. And the idea that her child, girl or boy, might become someone's joy, as Peter had been to so many, gave me peace.

With four weeks to go, Shannon was diagnosed with acute preeclampsia, a serious condition most often experienced in the last trimester of first pregnancies. As a precaution, she was hospitalized for complete bed rest until delivery. This became, in essence, a forced seclusion. While its precise cause is unknown, current thinking is that preeclampsia is an autoimmune response; the mother's body suddenly becomes "allergic" to the developing child. This reaction triggers the release of chemicals that can raise the mother's blood pressure to dangerously high levels, which can then damage blood vessels in the placenta (the organ that transfers oxygen and nutrients from the mother's blood to the baby's) and possibly lead to seizures and premature birth. When one of my sisters, meaning well, dropped by to visit,

Shannon's high blood pressure shot up practically two floors. Thereafter, she was allowed no visitors, save for one or two close friends, an arrangement not unlike living behind the grille at a Carmelite monastery. The nurses on staff, like a community of nuns, took care of her. And at the end of June, Shannon delivered a healthy baby boy, whom she privately named Daniel, though this was never entered into any paperwork. She had four days with him in the hospital before a nurse took him away. She never saw his new mother. My sister prayed that her child would have a good life.

Shannon took time off to heal and rest, scarcely venturing from her studio apartment. Two months after the delivery, she finally accepted a girlfriend's invitation to go out for an evening, her first time out in forever. Before leaving her place that night, Shannon told herself, *Tonight I'm going to meet someone I will never give away.* And she did. They've been together ever since and were married twelve years ago. His name is Daniel.

When I stayed with the two of them in their bright, new Seattle home last year, I noticed a small photo on their dresser of a dark-haired infant, a tiny wrinkled wonder in a bassinet. Shannon stepped closer to me as I picked up the frame. It took me a moment to make the connection: "Is this Daniel?"

"Yeah," she said, beaming. She'd put it out for a special occasion: The following day would be his birthday. This photo had been taken in the hospital. "He's going to be fourteen tomorrow. Fourteen—can you believe it?" Her eyes misting, I wrapped her in a hug.

We continued talking downstairs in the kitchen. "Anytime I look back on that situation," she said into the steam of her coffee, "I know that I made the right choice. But it was heartbreaking." Understandably, his birthday has always been rough for her, as are the four days that follow it, the span between the delivery date and the finalized adoption. Her son may feel something similar. Psychologist Nancy Newton Verrier, in her book *The Primal Wound,* explains that this is a recognized phenomenon among children adopted at birth or shortly after: "There seems to be a memory built into the psyche and cells, an anniversary reaction (also felt by the birth mother), which sends many adoptees into despair around their birthdays." Rather than celebrating a birthday like other kids, they may experience for several days the pain of having been relinquished, feelings formed long before their capacity to remember or understand them. Furthermore, Verrier notes, the emotions that well up on a birthday will often lead an older adoptee to wonder about his or her birth mother: *Is she thinking about me today?*

In years past, Shannon never "sensed" her son thinking about her, but that's changed dramatically, as if a psychic intercom has been switched on. "Now that he's a teenager, I'm sure he has an awareness of me. I'm sure his mom has told him about me, and I just feel him out there—here—someplace. I'm much more aware of his presence as he becomes more aware of mine."

Her talking so openly about her child is new. Only in recent years has she given herself permission to do so. The buffers around her emotions have eroded, it seems to me, as she's watched her nephews—our two sisters' sons in particular—going through puberty. Every time she sees Sam and Dylan, they appear to be an inch taller, a little huskier, more mature. Naturally, she can't help thinking, *What kind of man is my boy becoming?*

Shannon would welcome meeting him one day, as she indicated on the final adoption papers. However, she would never initiate a search for him, and if he chooses never to seek her out, she can accept that. Intuition tells her that he will, though, when the time is right for him. I hope it happens. I would love to meet him, too.

"I'm really curious about his voice, about hearing his voice," she admitted to me, smiling.

"It's probably deepening right about now," I pointed out. "Or cracking."

We both laughed, and I couldn't help but think of Shannon at that transitional age. When she was a scared young girl, whispering her secrets to me as we hid in the yellow bathroom, blood was so frightening to her. But today it carries such a different meaning. Now blood is a knock at the door, her son coming back to find her.

Vital Staining

WHETHER OR NOT THE NEED IS GENUINE, TRACKING down medical history is often an adult adoptee's stated rationale for instigating a search for his or her biological parents. This explanation provides an intellectual context for what's likely to be an emotionally raw experience for all involved. A woman considering having a child may first wish to learn if serious illness runs in her family—instances of childhood cancers, for example, or blood disorders such as hemophilia. Is there need to worry? In another case the search may be vital but unwanted: A middle-aged man, content to leave the past alone, must nevertheless locate a compatible organ or bone marrow donor. For others, the reason to search may seem silly when put into words—to find a relative who has, say, the same catsup-colored hair; finally, what made you stand out all these years helps you fit in. Despite the spoken purpose, however, adoption experts say that a search is usually driven by a deeper yearning. No matter how good one's health, how blissful one's upbringing, nothing may quiet the desire to know the people to whom you're related by birth, your true blood.

As I see it, the successful mapping of the human genome has done horrendous harm to the romantic notion of blood kin, a phrase that first entered the English language during the early Middle Ages. Medieval doctors believed that the act of conception was a mixing of "pure blood" from both the mother and father. This rarefied fluid originated in the heart and was carried by semen, a substance thought to be contributed by both sexes. This feeds into the emotionally satisfying idea that bloodlines are blood-made, ceaseless crimson tributaries reaching back thousands of generations. Even in the intimacy of the womb, however, circulatory systems are independent; blood does not cross over from mother to child. It is genetics that trickle down generation to generation, determining everything from hair color to blood type to a predisposition for certain illnesses. That's all well and good, but cold. The fact that I share DNA with my great-grandfather William, for whom I was named, does not have the same ring as hearing that we share the same Irish blood. Blood is tactile, warm, we are bathed in it at birth, whereas a spiral of DNA is clinical, invisible to the naked eye, the proof of something denied—a suspect's guilt, or paternity.

What we commonly learn as children is that we're part of a

family tree, each of us related to a number of great-great-long-dead people through a lattice of births and marriages. Steve's parents have spent the last several years digging into his family's roots, a project they've pursued with a degree of passion that might even make them honorary Mormons. Six months ago Millie and Ted sent us the product of their detective work, a computer disk with records accounting for six generations on both sides of the Atlantic. Where I thought the screen would sprout in a grand visual, all branches and leaves, instead a single name popped up, the current youngest member of the Byrne clan. By entering a number command, we could advance person by person, jumping back limb to limb, just a generation at a time. Once we got to Steve, we found me, listed as his partner. I should not have been so surprised to be included; his folks have always treated me like another son.

The program Ted had whipped up was deliberately simple; it boiled down lives to beginning and end dates with occasional footnotes—so-and-so had died of such-and-such, for example, or this cluster of family had immigrated from thereabouts to hereabouts at around this time. We poked around for other bits of information. We scanned the various branches for those births that had *oops* written all over them—the "Irish twins" born less than a year apart, the consecutive siblings separated by a decade or more. Steve and I also gave his family tree the gay inspection, looking for the curiously unwed, those "confirmed bachelors" and bachelorettes who might have had secret lives, secret families. Steve proudly pointed out several individuals around whom lavender suspicions had arisen. As we headed back in time, it did my heart good to see that so many of his ancestors had lived to ripe old ages, ninety and more. That he has such genetics can only help. Going through the document was continually unnerving in one respect, however. Between entering a command and the results appearing, the screen would go black, a disconcerting two-beat delay during which my mind would speed to worst-case scenarios: The program had been corrupted, a whole generation deleted.

The closest thing I have to a family tree is a collection of old address books, each one documenting a period of my life over the past two decades. They contain not just names but evocations of places, households—and also of the swath cut by AIDS. They are pieces of evidence, books I could never part with— proof of lives made, of family created then wiped away.

Of course, it's a rare person nowadays whose family fits into a perfectly traditional structure. Most of us have something less like a single tree and more like a "family orchard," a concept introduced by adoption educator Joyce Maguire Pavao. Whether you are adopted or a foster child or come from a blended home of multiple marriages—no matter how unconventional your household may be—Pavao's model acknowledges that your true family is often tied not just by blood or law but by circumstance and choice as well. It is this orchard that nurtures, feeds, and shelters those with whom you've found genuine kinship. Now, granted, a gay gym may seem like the last place for such an orchard to have thrived. But for twenty years, one did.

WHEN I GOT WORD OF THE CLOSING OF THE LEGENDARY SAN FRANcisco gym Muscle System on Hayes Street early last year, it was like learning an old friend had died and wondering, *Is it too late to pay my respects?* I hadn't worked out there in five years, not since I'd deserted it for a shiny new club that had opened near my home, but I regretted not having been around for the gym's final days. After getting the news, Steve and I made a trip down to Hayes Street to view its remains.

A café that shared the building remained open, so it was possible to stand inside the foyer and peer into the vacant space. It looked as though a tsunami had hit, flooding the gym and sweeping away all the weights. Left behind were the scattered skeletons of a few broken weight machines and tanning beds. The jade-green carpeting had been torn out (the very idea of a tastefully carpeted gym suddenly seemed like the epitome of gayness), exposing the raw concrete beneath. Prevented from going farther in by a wooden gate, Steve and I stepped up to the railing that overlooked the lower level and leaned over, as if on the prow of a ghost ship. All we could see in the shadows below were garbage cans where the stationary bikes used to be parked. One thing before us, though, remained unbroken and unchanged: the enormous wood-framed mirrors—covering every wall, floor to ceiling. Straight ahead, we could see our reflections in the wall opposite. Steve said, "We look farther away than we actually are."

I had joined Muscle System right after arriving in the city, even before I'd found a job and despite living nowhere nearby. At the time, it was *the* place to work out. It had such a mystique that Armistead Maupin wrote about it in his *Tales of the City* series. Every beautiful man in San Francisco had a membership to this gym, it was said. Luckily, I later met one there: Steve, who'd moved here from Illinois in 1987. Muscle System functioned as the heart of the community, even though it was located a good mile from the Castro district.

Monday evenings, after work, was Muscle System at its crazy best—150 guys, popping out of muscle T's, pumped. Within the human form, blood, it's been said, moves in figure-eights—from heart to body to heart to lungs; to heart to body to heart to lungs—circulating oxygen, nutrients, and heat, in endless loops. Exercise, of course, revs the cycle. By seven o'clock on a cold winter night, the furnace of bodies would raise the gym's temperature at least ten degrees. The street-front windows would steam over, and the place throbbed with endorphins and testosterone. At times, working out at Muscle System was more like being at a club: The towel boys behind the front desk danced as they fed the sound system; the floor teemed with all kinds of men—pups and bears and daddies; and guys fresh from the tanning beds vogued along the runway overhead. But moments like these, which seemed to recapture something we knew we'd lost, the innocence of pre-AIDS San Francisco, lasted about as long as one good song.

The impact of AIDS on the larger community could be seen in microcosm at Muscle System, where night after night we all came together, the grizzled veterans and the fresh-faced arrivals to the city. At the front desk, notes taped to the counter announced memorial services for fellow gym members and employees who had died. The notes often appeared before the obituaries were published in the local gay weekly, the *Bay Area Reporter*. I remember one for Mark, a congenial thirty-two-year-old southerner who, nearly every night for years, made a grand entrance after work. Although I never knew him well, I noticed when Mark was there, and his absence if he wasn't. Always arriving impeccably dressed in a suit, tie, and full-length camel-hair coat, with briefcase and gym bag in hand, he would throw a towel across his shoulder and sail to the locker room, waving "Halloo, halloo" to everyone in his path, adding each man's name if he remembered it.

When I'd last seen him about six weeks earlier, he appeared to have lost fifteen pounds. He actually looked good, his face as chiseled as Montgomery Clift's. I never saw him again. How could he have gone so quickly? It was as if one night at the gym—working out, as always, in fluorescent, confetti-colored bike shorts and a tank top—Mark had simply walked right through the mirrors and disappeared.

I remember thinking in the days that followed, Now his ghost is here, behind these mirrors, together with all of the city's most beautiful dead. They watch us as we stare at ourselves, all lined up, clutching the weights.

"COME ON, LET'S GO IN," I SAID TO STEVE. THE GATE'S SIMPLE LATCH gave way with a press of my finger.

The girl behind the café's counter called out, "Um, 'scuse me, you're not supposed to go back there!"

But we're members here, I wanted to say, *lifetime members.* "Don't worry, we won't touch anything, we'll just be a minute," I said. Steve and I certainly couldn't have made the place look worse, more vandalized, than it did.

Careful steps took us along the familiar path back to the locker area. Although the sauna had been deconstructed, the banks of lockers stood unchanged—row after row, like two hundred metal time capsules. I half expected to find them filled with members' clothes. One after another, we opened every locker and collected what little we found: a penny, a key, a video rental card. Inside every door, though, was a sticker, which he and I had surreptitiously put there, visit after visit after visit, a dozen years back. We laughed in amazement—of all things to have survived! We had slapped them up in defiance of the owners' prohibition against distributing AIDS educational materials on the premises. BE HERE FOR THE CURE, the stickers read, the words above a luminous painting of the globe. One was scribbled over and read, *ACT UP* FOR THE CURE! Another, probably scrawled more recently, said, WE'RE STILL WAITING.

At the time of our stickering, both Steve and I worked just a couple of blocks away at the San Francisco AIDS Foundation, he on the hotline, me in the education department. I had been charged with creating a media campaign to promote the then novel strategy of seeking out early medical care if you were HIV positive and asymptomatic and had come up with the "Be Here for the Cure" theme. The idea behind it was that the sooner you got tested, saw a doctor, and started treatment, the better were your chances of longevity. The message spread throughout the Bay Area on T-shirts, buttons, stickers, posters, and billboards, in treatment packets, ads, and PSAs, in multiple languages. What I'd learned from extensive interviews and focus groups was that, despite the number of AIDS deaths (twenty-six thousand in California by the end of 1991), many in the gay community still had hope—not always for themselves, but always for the next generation. I know I felt it. If asked back then, I'd have said with certainty that the cure would eventually arrive in some form of magic bullet, perhaps as a wonder pill or a single shot in the arm, a so-called therapeutic vaccine. The very notion of

a magic bullet, entrenched in the lexicon of illness, would've required no further explanation, no translation. Thinking about it now, though, the phrase strikes me as one of those familiar word pairings that seems more a product of free association than of deliberate coupling, like *friendly fire* or *drug cocktail,* nonsense made meaningful, light bent to gravity.

In fact, the term *magic bullet* was coined in 1908 by a brilliant fifty-four-year-old German scientist named Paul Ehrlich, who that same year received a Nobel Prize for, in the committee's words, his "immortal contribution to medical and biological research," work that laid the foundation for the emerging field of immunology. Today Ehrlich is probably best known for being the first scientist to propose using high-dose chemical compounds to destroy specific pathogens or cancerous cells—what's now called chemotherapy. Like "magic bullets," Ehrlich explained, such compounds would fly through the body, "straight onward, without deviation," and "find their target by themselves," causing no harm to surrounding tissue. The concept was radical because, up till then, chemical agents had been used principally to treat symptoms—fever, pain, sleeplessness—never to eradicate disease.

In the year following his Nobel Prize, Ehrlich did indeed create the world's first magic bullet when he invented an effective cure for the most hideous plague of his day, syphilis, the sexually transmitted, blood-borne disease that, for centuries, had been as stigmatizing as AIDS would one day become. He formulated an injectable arsenic-based drug that was later called Salvarsan by the German manufacturer. (Salvarsan would eventually be replaced by penicillin as the first-line syphilis treatment.) Ehrlich's original name for the medication was "606," for the simple fact that it was the six hundred sixth preparation he had tested, the number also quietly acknowledging his sheer persistence even after 605 failures. While the discovery made Ehrlich world famous, it also marked the beginning of a new set of difficulties, placing the scientist at the center of an ethical debate. From one camp, Ehrlich was vilified for his willingness to save "immoral" people who, some believed, deserved to die. And from another, he was held personally responsible for the drug's adverse side effects, including numerous fatalities, most of which resulted from doctors' errors—incorrect dosing and poor administering. In reality, the magic bullet was not magic for all. Produced as a powder that had to be carefully measured then dissolved in sterile water prior to each intravenous injection, Salvarsan was also difficult to manufacture. Ehrlich, in an effort to minimize the risks, had taken the extraordinary step of patenting Salvarsan (one of the world's first therapeutic drug patents), not for personal gain—in fact, he never directly profited from the drug—but to enforce a consistent quality in its production. What made this whirl of difficulties bearable, he later confessed, was the first postcard he received from a cured patient.

Twenty-five years after his death in 1915, the doctor's life story was dramatized—no, make that *melo*dramatized—in a Warner Bros. film, *Dr. Ehrlich's Magic Bullet,* starring Edward G. Robinson in the title role. Hollywood—and hindsight—treated him kindly. The 1940 movie, with an Academy Award–nominated screenplay by John Huston, is notable for being the first to address the taboo subject of syphilis. Ehrlich is presented as a selfless, courageous German Jewish doctor with an American accent to set him apart, I presume, from the vaguely anti-Semitic government bureaucrats, all of whom have heavy German accents. (This reflected the politics of World War II more than the reality he'd faced.) Robinson, best known for portraying gangsters, gives Ehrlich a saintly aura, culminating in the near apotheosis of his deathbed scene. Looking and sounding robust—though the funereal piano score leaves no doubt whatsoever that he's about to kick the bucket—Robinson-as-Ehrlich summons his scientific disciples to his bedside: "The magic bullet will cure thousands," he tells them. "But there can be no final victory over diseases of the body unless the diseases of the soul are also overcome." A pause and a benevolent smile as the master musters his final strength. "We must fight them in life as we fought syphilis in the laboratory. We must fight! Fight! We must never, never stop fighting!" His eyes close as violins swell above the piano, then bells ring as if signaling Ehrlich's entrance into heaven. And the screen goes black.

By all published accounts, Paul Ehrlich's real life did lend itself to this kind of glorification. Most glowing of these works is a memoir written by Martha Marquardt, who served as the doctor's devoted secretary during his final thirteen years. (Her loyalty to the man continued long past his death, as it turned out. At the risk of imprisonment, she smuggled Ehrlich's private letters, scientific papers, and original manuscripts out of Germany at the height of Hitler's regime, thus saving them from certain destruction.) In point of fact, Marquardt wrote two versions of her memoir: the original, a slim volume of reminiscences from 1924, and a substantially revised, English-language edition from 1951, incorporating the documents she'd rescued. In this latter work, the additional material allowed her to write more of a full-fledged biography. But there was another compelling reason for the new edition. As Marquardt noted in the preface, all but a few copies of the original were burned by the Nazis.

The moment she entered the lab first thing in the morning, she wrote, Ehrlich would nod courteously then start rattling off correspondence. Midsentence, though, he'd often abruptly halt, as if listening for something just out of range of human hearing, and then begin rummaging through the cork-stopped bottles atop his immense worktable. Not finding what he sought on top, he'd open the cupboards underneath. Here were still more bottles—"innumerable they seemed, filled with rare and precious chemical substances."

Marquardt recalled that the doctor might remain squatting for a good quarter of an hour, his knees pressed to his chest. The sound of his rifling was the tinkling of a tea service. He'd then pick up with both hands a particular bottle, turn it around and around, and smile as he read the label. With the grunts and groans of repositioning, he again stood. At such moments, she remembered, "All written work was forgotten for the time being and he would begin experimenting. Test tube after test tube was taken out of the little box near the Bunsen burner, and minute quantities of various chemical compounds were put into them, solutions were made and heated, alkalis and acids added. Now a delightful violet-blue resulted from the experiment, then it was a bright red; now green, then orange. If he found an interesting re-action he called out: *'Wonderful! Wonderful!'* and showed it to me as though I understood all about it."

In Marquardt's affectionate work as well as in the drier tomes of Ehrlich's fellow scientists, a telling quirk surfaces: The man loved colors. He was "emotionally swayed" by them, one gentle-man wrote. What spring is to a *parfumeur*, colors were to Paul Ehrlich. Though always a busy, busy man, Marquardt revealed, the doctor would still stop to extol the roar of yellows and reds in a bouquet of flowers. They "would make him quite ecstatic," Marquardt admitted. This quirk carried over into his work habits as well. He wrote daily notes to himself and his staff on precut squares of different-colored construction paper, using various colored pencils. He kept this stationery in his coat pocket and, while he rarely lost his temper, he'd become terribly agitated if his supply ran out. (A similar response arose regarding his stock of Havana cigars, one of which was ever present, a sixth digit on his right hand.) Though surrounded by colors, none were more eye-popping to Dr. Ehrlich than those produced by his chemicals—the pure blues of cobalt compounds, glowing like the core of a torch flame, the delicate sea greens of iron-infused solutions. Far more than a source of pleasure, though, color was the prism through which he viewed and attempted to unravel biological mysteries. Color is the thread that links his disparate scientific achievements.

Born in 1854 in a small village 150 miles southeast of Berlin (an area that's now part of Poland), Paul Ehrlich was the only son of prosperous Jewish parents who operated an inn. During his midteens, he pursued his keen interest in science under the watchful eye of a cousin on his mother's side, Carl Weigert, who was nine years older than Paul. Weigert, a famous pathologist, had discovered that aniline dyes—synthetic dyes developed in Germany around 1860 for use in the textile industry—were unexpectedly well suited for staining human and animal tissue. Rather than obscuring details, these intense dyes instead illuminated them, revealing contrast and texture, making microscopic specimens easier to analyze. Weigert introduced his cousin to this important advance, and Paul began experimenting on his own. In 1872 he went off to study medicine at the University of Breslau and, as was customary in those days, transferred to different schools every year to train with the finest teachers. At the University of Strasbourg, while under the tutelage of one of the great anatomists of the nineteenth century, Wilhelm Waldeyer, Paul invented a technique that took his cousin's discovery one step farther: "selective staining." Using a dye of his own formulation, he found that each cellular element in a tissue sample reacted differently to his staining and also displayed a different shade, thus permitting extraordinarily sharp definition—akin to what HDTV is to regular television, I imagine. With this method, Paul promptly made his first major discovery: the mast cell, a type of cell common in connective tissue.

Praise for his innovative staining was not unanimous, however. While completing his studies at the University of Leipzig, Paul lived at a boardinghouse whose owner would recall many years later that the young student often looked like a human drop cloth; his hands, face, and clothes were always spotted with inky stains. His bath towels and linens were equally blemished, and no amount of washing could remove the blight. Moreover, even the house's billiard table, upon which, for lack of flat surfaces, Paul had conducted experiments, was forever splotched with fuchsia, indigo, and lilac. Little could the proprietor have known then that her boarder's messes would lead to findings that would make a permanent impact on several branches of science. Ehrlich had not been content simply to observe the myriad cells of the human body—as dazzling as he found them—but sought to figure out *why* dyes became fixed in specimens (as they did in fabrics), plus why individual cellular parts reacted so differently to certain dyes. From controlled experiments, he concluded that, rather than a mere physical change, a specific chemical transformation was occurring within the cells. These investigations into the nature of staining were the subject of Ehrlich's 1878 doctoral thesis and also anticipated the dawning of a new field of biology, cytochemistry, the study of cell constituents. The thesis also contained the germ of a larger theory explaining how—in lay terms—disparate substances bind chemically, which would evolve over the next thirty years into his ideas about the formation of antibodies in human blood; his concept of a magic bullet; and, ultimately, the invention of the syphilis cure. But that's jumping ahead.

Already renowned for his histological staining, the newly minted doctor was invited in 1878 to join the staff of the prestigious Charité Hospital in Berlin, where he worked under the supervision of Friedrich von Frerichs, an esteemed clinician. Though Ehrlich had a full roster of patients, Dr. Frerichs recognized the young man's talent and energy and encouraged him to spend more time on original research. Ehrlich continued experimenting with aniline dyes but turned his attention from animal tissue to human blood, which, in a clinical setting, was readily available. The study of blood, too, was still fresh terrain. Although two centuries had passed since Antoni van Leeuwenhoek's discovery of red blood cells, progress in the field of hematology had been sluggish. From today's vantage point, it appears that Ehrlich was the right scientist at the right time and place, equipped with the right tools, to transform the entire discipline. Ensconced in his closet-sized hospital lab, the twenty-four-year-old quickly invented a simple method for preparing blood specimens. Well, simple for him. First he would take great care to spread a small drop of blood on a glass slide in the thinnest possible layer. He then allowed it to air-dry. Next, as he reported in a published paper, he heated the blood smear on a copper hot plate "for one or several hours" at 120 to 130 degrees Celsius, thereby fixing it and preserving the delicate cellular elements. Finally, he added one of his staining solutions. "Using such techniques," Ehrlich concluded, "one can obtain with most dyes very beautiful and elegant pictures."

Of great immediate significance were his "pictures" of white blood cells. Though a predecessor, the British microscopist William Hewson (1739–1774), had discovered white blood cells a hundred years earlier, he had supplied only sketchy details. (Medical historians now explain that white cells had been overlooked for so long not only because they're far outnumbered by red cells, which make up 45 percent of blood volume, with white just 1 percent, but also—for me, the more convincing reason—because they are almost transparent.) Hewson also correctly surmised that white cells serve an infection-fighting role, in concert with the lymphatic system. Confirming Hewson's theory, another British William, physician William Addison, demonstrated in 1843 that blood collected from an injured person's wound had far more white cells than blood taken from elsewhere in that person's body. Clearly, white cells were converging, but for exactly what purpose? Subsequent scientists, leading up to Ehrlich's generation, discerned that white cells were, indeed, the army of the blood—helping defend the body against bacteria, fungi, and viruses. But it was Paul Ehrlich who first identified its main soldiers. By using his technique of selective staining, he differentiated the two broad classes of white cells, leukocytes and lymphocytes, and discovered three of the five specific types of white cells now known. He found he could illuminate each of these three using a different kind of dye. The names Ehrlich gave the cells were mini tributes to the dyes themselves: the eosinophil, which stained red from the eosin dye; the basophil, blue from a base dye; and the neutrophil, a pinkish color from a neutral dye.

Ehrlich's heat-fixed blood-staining technique soon became standard practice, helping to usher the science of hematology into the modern age. Adopting his methodology, other scientists joined Ehrlich in making finer and finer distinctions about the behavior and function of both red and white blood cells, particularly in diseases such as anemia (characterized by a lack of hemoglobin) and leukemia (an overabundance of white cells). Likewise, quantitative blood counts were now possible and allowed for quick, accurate diagnoses of life-threatening conditions. The modern-day routine blood test called a CBC (complete blood count) is a direct descendant of Ehrlich's innovation.

For his accomplishments, Ehrlich is now often lauded as the "father of hematology," a fact that calls for a small digression. Medical historians, I've noticed, demonstrate an almost comical penchant for assigning paternity to branches of science even if the field is already well populated with fathers. In my studies, Ehrlich is the fifth father of hematology I've come across, but this in no way minimizes his contribution. Among these patriarchs, though, Ehrlich stood out in one respect: He didn't dote. While it is not unusual for a scientist to devote his or her life's work to one specialty, Ehrlich tended to make discoveries and then, as I see it, abruptly move on. But he described it with greater eloquence: "One must not stay in a field of work until the crops are completely brought in, but leave still some part of the harvest for the others." He moved on, again and again, to great success. Historians, in fact, also remember him as the father of histology, immunology, experimental oncology, and the aforementioned chemotherapy and cytochemistry.

Not all of Paul Ehrlich's pursuits were academic, however. Five years into his tenure at the Charité Hospital, the then twenty-nine-year-old confronted a malady for which a cure is not always welcomed: lovesickness. The object of his infatuation was Hedwig Pinkus, a petite beauty ten years younger than him, the daughter of a prominent family from his hometown of Silesia. Though he squeezed in trips to visit her, Paul, for the most part, courted Hedwig through daily letters. An excerpt from one reveals a charmingly besotted man of science: "Although I am not allowed, dear Hedwig, to delight in your presence," he wrote on March 2, 1883, "the thought of you has not left me for one moment. I must confess I am of no use at all for anything else, but I cannot help it. I have never been as unscientific as now. . . . My microscope is rusting, my beloved dyes are becoming moldy, the laboratory is collecting dust, and the [lab] animal keeper is shaking his head in disbelief." Married in August 1883, the Ehrlichs settled down in Berlin; within a year the two became three with the birth of their daughter Stefanie.

A year before the birth of a second daughter, Marianne,

Ehrlich's life was shattered in March 1885 with the suicide of his revered mentor and ally, Dr. Frerichs. The pain of this loss was compounded by the frosty relationship he developed with his new boss, Frerichs's successor, Dr. Carl Gerhardt. A stern taskmaster, Gerhardt decreed that Ehrlich must now devote the entirety of his time to patients. This abruption of his research could not have come at a worse moment. Ehrlich, using a dye called methylene blue in experiments with frogs, had just succeeded in staining *living* nerve tissue. This was a major technical breakthrough, for he, like other scientists, had always worked with inert samples of tissue and blood. Vital staining, as this was called, allowed him to begin examining the effect of chemical compounds on live cells, the important next step toward the crowning achievements of his career (a subject continued in the following chapter). But for now, this work would have to be

A tender moment between Paul Ehrlich (portrayed by an almost unrecognizable Edward G. Robinson) and his wife, Hedwig (Ruth Gordon), in a scene from the 1940 Warner Bros. film Dr. Ehrlich's Magic Bullet

shelved. Unable to pursue what he most loved, Ehrlich, miserable under Gerhardt's command, finally resigned his post at the Charité after two years.

Complicating his decision—or perhaps, in a certain way, simplifying it—he'd developed a persistent cough from which he'd been unable to recover. Shortly after his resignation, he discovered evidence of tuberculosis in his own sputum (likely contracted from patients' TB cultures), a finding that was sadly ironic because, just a few years earlier, he had invented the heat-dried staining method employed to make such a diagnosis. With a firm nudge from his worried wife, Ehrlich decided then not only to put his career on hold but also to take the radical step of leaving Germany altogether. At age thirty-four, he formalized plans to move his young family to Egypt, where he hoped he'd recuperate more quickly in the warm, dry climate.

Many years later Paul Ehrlich looked back on these final weeks in Berlin. During his lowest moments while working under Dr. Gerhardt, he recalled, those times when he felt bitter, dejected, he'd sneak away to his dusty laboratory, open the dye cupboard, and drink in the bright colors. He'd remind himself, "These here are my friends, who will never desert me."

Detectable

BLOOD NEVER SLEEPS. EVEN WHEN WE'RE DEAD TO THE world, sad and torpid lumps under the covers of a sickbed, our blood is mounting its most vigorous defense. Here's the drill: Approximately thirty minutes after we launch into sleep, the killers come out in full force—the "killer" T cells. Killer cells are lymphocytes, one of the five broader varieties of white blood cells. Their territory is our bloodstream and the connecting lymphatic tissue. Killers are created for a single purpose: to destroy foreign agents—viruses, bacteria, toxins. When a killer cell comes upon a virus, for example, it gloms onto it, then secretes proteins that riddle the germ like Swiss cheese, slaying it—mission accomplished—but at the same time sacrificing itself. Killers are most numerous at night, though they operate around the clock, as do their fellow T cells, the "helpers" and "suppressors," which also perform crucial roles in our defense. (All three T cells take their *T* from the thymus, the butterfly-shaped gland located between the heart and breastbone, where they mature.) The other kind of lymphocyte, the B cells, which develop in the bone marrow, also emerge during deep sleep. And they, too, exist to make mincemeat of microbial bodies, but their methods are less direct. B cells produce preprogrammed weapons called antibodies, which head off into the blood to carry out their orders.

Our blood works not only to destroy the uninvited, but also to repair. In sleep, our circulatory system is infused with growth hormone, a product of the pituitary gland and essential in helping rebuild damaged tissue. Growth hormone also rouses additional infection-fighting substances called cytokines, which, like a densely worded paragraph, can make us sleepy. It's an extraordinary give-and-take. As sleep bolsters our immune system, our immune system bolsters sleep.

So, as it turns out, our moms were right all along.

Get back in bed this instant. Or, *All you need is a good night's sleep.*

These refrains of countless generations of mothers are grounded not just in the clear messages our bodies send us but also in sound science. And while the mere mention of bed rest as a cure-all inevitably conjures up for me images from childhood, complete with the Vicks VapoRub and those spoonfuls of "grape-flavored" god-awfulness, the fact is, round-the-clock rest as a clinically proven treatment for sickness was first prescribed 150 years ago—the brainchild of a scientist from Paul and

Hedwig Ehrlich's hometown. Hermann Brehmer, a botanist from Silesia, contracted TB during the late 1840s and moved to the Himalayas to live out his final days. The young man's prognosis was grim. TB, also called consumption, was almost never survivable and, as we now know, this bacterial infection has existed since prehistory; the royal mummies of ancient Egypt show clear evidence of its ravages. Between 1700 and 1900, according to historians, an estimated one billion people died of the disease. Hermann Brehmer did not expect to be an exception. To his great surprise, however, the fresh mountain air and abundant bed rest worked wonders, and he fully recovered. (What Brehmer's unplanned regimen had done was deprive the bacteria of the conditions they needed to thrive, an immunologist today would explain, thus giving his immune system the edge it needed to fight back.) Following his return to Germany, Brehmer published in 1854 the banner-titled book *Tuberculosis Is a Curable Disease,* in which he espoused his TB "rest cure." That same year he opened the world's first tuberculosis sanatorium, the prototype for thousands built in Europe and the United States in the decades to follow. (By the 1940s these facilities had run their course, made obsolete by the widespread availability of antibiotics.) A mainstay of sanatoriums was the sleeping porch, where patients could rest, soak up sunshine, and "take the airs." Of course, in true Michelin guide style, sanatoriums ran the full range of stars, from squalid public institutions to luxurious resorts for the well-to-do. In fact, when money was less of a concern, patients such as Paul Ehrlich could even make a vacation of their recovery, just as long as they made the commitment to follow doctor's orders.

Now, one would think that Paul would've taken to forced rest like a cat to bathwater. Even Hedwig expected her thirty-four-year-old husband to go partway 'round the bend. In their five years together, she'd scarcely ever seen him take a day off. And yet, even before the Ehrlichs had reached their final destination of Egypt, the good doctor was showing early promise. The couple and their two daughters had first stopped over at a lakeside spa near Venice, and Hedwig, it was later reported, admitted her wonderment at how quickly Paul was adjusting to full-time R and R. "People always think I'm a hard worker," he remarked at the time, "but they're wrong. I can be as lazy as a giant snake."

True enough. As Hedwig well knew, his favorite idle pastime had always been getting lost in books. So as not to disturb her sleep, he'd even divided their Berlin bedroom with a black curtain, which he'd draw and then read behind into the early morning. A perusal of his bookshelves would've revealed the breadth of his interests, from erudite leanings—including Greek classics and the latest works by contemporaries such as Friedrich Nietzsche, whose hymn-like verses Paul could recite by heart—to the other extreme, his great love, detective stories.

Every biographer summing up Ehrlich's life mentions his passion for detective fiction. Martha Marquardt, for instance, revealed that Saturdays at the lab were made sacred by the arrival of the latest issue of the doctor's favorite crime magazine, with, as she described with an implied *tsk-tsk,* "its cover showing the most lurid pictures of murder." This weekly serial magazine was probably akin to the American "pulps" that became popular in the early 1900s, such gritty treats as *Detective Story Magazine, Black Mask,* and *The Shadow.* Though they were called pulps because of the cheap, wood-specked paper they were printed on, the stories were sensational and soaked in intrigue. Paul would devour the new issue that same night, Marquardt reported, and it never failed to distract the doctor from his true-life problems.

Ehrlich was also a huge admirer of Sir Arthur Conan Doyle, a signed portrait of whom held pride of place on the wall of his study. He owned copies of many of his books, several of which had been personally inscribed by the Scottish physician-turned-author. As to the when and wherefore of the first "meeting" between Ehrlich and Sir Arthur's most esteemed creation, the inimitable Sherlock Holmes, my sources do not say. But if one considers that the first Holmes novel, *A Study in Scarlet* (1887), was published just a few months before Ehrlich began his convalescence, it's not too great a stretch to imagine he brought along a copy of the new whodunit.

"Voilà, hemoglobin."

While *A Study in Scarlet* is most memorable for presenting the first meeting between Holmes and Dr. John Watson, I take particular notice of what immediately follows this historic handshake. Holmes, in his own estimation, has just made an utterly brilliant discovery about bloodstains. He seizes Watson by the coat sleeve and tugs him into the spacious laboratory to demonstrate said brilliance. Given that arrests were often made long after the commission of a violent act, the detective explains, it had theretofore been difficult for the London police to prove that incriminating stains found on a suspect's clothing were blood rather than, say, fruit or rust stains. But no longer, as Holmes shows. He pricks his own finger with a needle, draws some blood into a pipette, and stirs a drop into a liter of water. Of course, all evidence of scarlet disappears. But wait. Holmes, re-creating an actual forensics innovation of the time, crushes a few white crystals into the water, followed by several drops of a transparent fluid. In an instant, the liquid takes on a dull mahogany color, and a brownish precipitate collects at the bottom. Voilà, hemoglobin. Holmes is so delighted with himself he'd be patting himself on the back were his hands not occupied with the experiment.

When Dr. Ehrlich did read *A Study in Scarlet,* I can't help but wonder if he noted the characteristics he and Sherlock Holmes shared: how both men's hands, to borrow Watson's words, were "invariably blotted with ink and stained with chemicals"; how, although they both brought a broad background in the sciences to whatever subject was at hand, each possessed an enthusiastic "knowledge of sensational literature"; and how both men incessantly smoked strong tobacco (no, not even TB could make Ehrlich give up cigars). There may have even been aspects of this character's life that Ehrlich dreamed of having for himself—the unquestioned autonomy, for instance; the instant respect; and, perhaps above all, that gloriously spacious laboratory.

Brushing aside all speculation now, the fact is, when the Ehrlichs returned to Berlin in the spring of 1889, Paul was a new man—free of TB, hale and hearty, raring to jump back into full-time work. Just one snag: Nobody was hiring. Though far from his dream scenario, he made the best of the situation. With the financial backing of his father-in-law, the thirty-five-year-old opened his own research laboratory, which may sound more glamorous than it was. In truth, it was a rented apartment, close to where he and his family lived. And his staff rounded out to a whopping one, a valet named Fritz, although Paul's nephews, Felix and Georg, did pitch in now and then. Picking up where he'd left off the year before, Ehrlich resumed his experiments with vital staining and began creating new histological dyes. He christened Stieglitz blue and Lutzow blue for nearby streets. A more informal term—*exploders*—arose to describe a common mishap: the bursting of the dye-filled glass flasks that were heated on the apartment's kitchen stove, leaving indigo spattered about the room.

Elsewhere in the small apartment, Ehrlich launched into what was for him a new line of research, work that was related to a larger "hot theory" being addressed by scientists in Berlin, as elsewhere: that all infectious diseases were caused by toxins, a by-product of foreign microorganisms. Scientists had just concluded that this was the case in diphtheria, for instance; the diphtheria bacterium secreted a toxic substance that attacked the walls of the throat, producing the blockage that left its victims, mostly children, choking to death. Soon after, it was found that a toxin was also the culprit in tetanus. Scientists then cast a suspicious eye on TB (although, eventually, no toxin was implicated). Ehrlich, a man who "approached research like a detective on a trail," as the distinguished American hematologist Maxwell Wintrobe wrote in 1980, began focusing on one small aspect of the whole. He mounted his own quantitative study of a toxin, but rather than something infectious, Ehrlich chose something addictive: cocaine.

At the time, cocaine was legal and readily available, whether in pure form from a pharmacist or, as was the case in Anytown, USA, at the corner drugstore as the extra little kick in a glass of Coca-Cola. Coincidentally, the second Sherlock Holmes story, *The Sign of Four* (1890), had just been published, and it opened with Holmes casually injecting himself with a syringe of cocaine, the influence of which, he confessed to Watson, he found "transcendentally stimulating and clarifying." Its popularity notwithstanding, Ehrlich knew that at certain levels cocaine had toxic effects. But what levels caused what effects? For answers, Ehrlich enlisted mice as his guinea pigs. Rather than injecting the cocaine into their bloodstreams, he instead found it easier and safer to feed it to them. He soaked biscuits with varied but precise quantities of a cocaine solution. Satisfied with this methodology, Ehrlich shifted to a series of experiments using a much deadlier plant derivative, the toxin ricin. Derived from castor plant beans, ricin is more potent than cobra venom, even in minuscule amounts. Today it is regarded as one of the most dangerous weapons of bioterrorism.

Although Ehrlich would end up with a lot of dead mice, he eventually produced survivors that were immune to not just normally lethal amounts of ricin but also doses hundreds of times stronger. Within the bloodstream of these supermice, Ehrlich had triggered circulating "antitoxins" (a type of antibody) that would "paralyze" the poison the next time the mice ingested it. In short, they'd been vaccinated. With this, Ehrlich was not, however, introducing into the world a new concept. A hundred years earlier one of his scientific heroes, the British physician Edward Jenner, had demonstrated an effective but far cruder instance of induced immunity. He'd found that a human being deliberately exposed to a mild form of smallpox, by way of scrapings from sores, would survive exposure to the deadly form of the disease. Although the how and why were unclear, Jenner had proven, figuratively speaking, that an umbrella against drizzle could be as effective in a downpour. In turning to this puzzle a century later, Ehrlich brought to the table his own specialty: great scientific rigor. With his ricin experiments, he developed a scrupulous methodology, juggling multiple factors. As a result, he knew exactly how much of the toxin was required to elicit immunity, according to a specific dosing schedule over a certain number of days. Likewise, he knew what amounts were too little or more than necessary. When his supermice had offspring, Ehrlich discovered something else of significance. The antitoxins were being passed on by the mother, from placenta to fetus as well as through suckling—textbook examples of passive immunization, in which an individual receives antibodies from another. (In active immunization, by contrast, protective antibodies are generated by one's own immune system.)

Ehrlich's methods and precision caught the attention of his peers, one of whom, fellow Berliner Dr. Emil von Behring, had just made his own startling discovery regarding passive immunity. In experiments completed in 1890, Behring found that if he removed the serum (the plasma without the blood cells and clotting elements) from an animal that had been successfully immunized against diphtheria, and then injected it into a second animal, that animal would also be immune. Serum from the injected animal would, in turn, protect other animals. Taking the next step, however— creating a diphtheria antitoxin to protect human beings—had proved troublesome. Behring wisely enlisted Ehrlich's help in developing a safe, effective protocol. Ultimately, full-scale production of the lifesaving diphtheria treatment began in November 1894.

Paul Ehrlich in his laboratory

From there, a five-year jump in time finds a world-recognized Paul Ehrlich as the head of his own institute, the newly established Royal Institute of Experimental Therapy, located in Frankfurt—a long distance, both geographically and professionally, from his cramped quarters in Berlin. The institute had been designed to Ehrlich's every specification, with multiple laboratories, a library, and ample space for a top-notch staff plus countless lab animals, all housed within a grand four-story building. Ehrlich oversaw a broad range of work whose scope was comparable, for its time, to, say, the United States' National Institutes of Health combined with the Food and Drug Administration. While its opening ceremony in early November 1899 was a splendid public affair, attended by scientists, journalists, politicians, and citizenry, for Dr. Ehrlich personally a much more prestigious, albeit quieter, event would take place four months later.

It is March 22, 1900, and the forty-six-year-old Paul Ehrlich stands before the Royal Society of London—the exclusive scientific association that counts Antoni van Leeuwenhoek and Sir Isaac Newton among its past members. When he speaks of the great privilege it is to be here, this is no mere nicety. He has been invited to this first gathering of the new century to deliver the keynote address, a lecture titled "On Immunity with Special Reference to Cell Life." He does not disappoint. In this now-legendary speech Dr. Ehrlich elaborates for the first time his "side-chain theory" of immunity, which provides a full accounting of the blood's ability to protect the body from foreign invaders. Drawing upon the work of peers as well as his experience with ricin and diphtheria, Ehrlich explains that blood cells have on their surfaces ready-made receptor molecules, or "side chains," that link or bind chemically with certain invading toxin molecules. (He had borrowed the term *side chain* from organic chemistry; it was widely believed that side chains were, like docking ports, the means by which cells took in nourishment from free-floating food particles.) Long story short, this binding neutralizes the toxins. It also triggers production of excess side chains, which are released into the blood as circulating antitoxins to fight the same toxin in the future.

Medical historians today distill Ehrlich's presentation to three main points, two of which were correct and one that was wrong but forgivable. He was right in theorizing that blood cells have the capacity to form antibodies even before a particular antigen has entered the body. Also right was his conception of these antibodies as, in essence, locks waiting for the right keys; and the related notion that, once a lock was activated, the production of many more antibodies was stimulated. Ehrlich was mistaken, though, in believing that *all* cells could produce such antibodies; in fact, only B lymphocytes can.

Historians also agree that Ehrlich's argument was not just sound but very convincing. Adding to the impact of his spoken word was a series of provocative drawings. Now, it should be noted that the use of visual aids in a lecture to the Royal Society was not at all unusual, but his were unique for being of imagined constructs, renderings of the theoretical goings-on in the blood. Even though the best microscopes of the day did not allow Paul Ehrlich to see this activity, in his mind's eye the images were clear. And were now on display. A sequence unfolded: First a standard cell was shown—a light-colored, spongy moon erupting with what looked like sweat beads off the brow of a comic-strip character. These were Ehrlich's side chains, which frankly did not in any way appear chain-like. Next, some of the sweat beads were gripped by toxins, the villainous elements, which were horned and black. Others then broke free—the heroic antitoxins—and, now resembling lithe, silvery minnows, swam off into the blood.

Ehrlich, aware that not everyone in the audience would share his certainty, cautioned that the forms and shapes in his diagrams should be considered as "purely arbitrary." At best, they were simply an educated guess. Some scientists did not take this caveat to heart, however. In the weeks and months to follow, critics griped that his ridiculous "cartoons" had conveyed more of a conclusion than a possibility. His chief detractor dubbed them a "puerile graphical representation." If the criticism was meant to elicit a retraction of some sort, it didn't work. In fact, whenever Ehrlich subsequently spoke about his side-chain theory, he would take the opportunity to illustrate it. Stories abounded of how Dr. Ehrlich, even in casual conversation with colleagues, would scribble out drawings of the minuscule players of his theory on whatever blank surface was available. When no paper was handy, he'd opt for, say, his hostess's tablecloth, a listener's shirt cuff, or the sole of his shoe. If necessary, he'd roll back the carpet, then use chalk on the floorboards. And once, over dinner, in a performance I'm sad to have missed, he storyboarded his entire molecular drama on fifty postcards, an indulgent waiter having kept the doctor in steady supply.

Martha Marquardt, who entered his employ in 1902, adored this quality in her boss. "When his mind was entirely filled with a certain idea," she wrote, he spoke of it with animation and in a great gallop of words. "He perceived the idea as if it had a physical existence," and he always wanted visitors to see it along with him. To make sure a person was keeping up, Dr. Ehrlich might tap him or her "lightly on the arm or chest with the point of a coloured pencil, with a test tube, a cigar or his thick-rimmed spectacles which he frequently took off and swung about. . . ." Drawing to a finish, "he stood with his head pushed forward a little, his gentle face upraised," and "looked penetratingly at the other person with his big bright eyes." *Do you see what I see?*

DR. EHRLICH NEVER ACTUALLY GOT TO VIEW THE DRAMA WITHIN THE blood. But what was to him the most likely scenario involving the most likely suspects can now be clearly photographed with an electron microscope, the same technology used to produce those ugly mug shots of minute insects, with their bulbous compound eyes. The electron microscope, thousands of times stronger than the traditional compound microscope, has also captured images of a tinier but more horrific bug: HIV, the virus that does its damage to the immune system by hijacking helper T cells and forcing them to churn out as many copies of itself as possible, a process that kills the cells. I remember the odd sense of relief I felt upon first seeing the micrograph of HIV on the cover of *Time* magazine, dated August 12, 1985, one month after I'd moved from Seattle to San Francisco's Castro district, ground zero of the epidemic. It showed the virus, magnified 135,000 times, attacking a T cell, according to the caption, although the grayish clump looked more like something pulled from a vacuum cleaner bag. *There's the culprit,* I thought, staring at the black-and-white photo. *Now we just need to annihilate it.*

I've since seen many similar images, some taken at magnifications three times as powerful. Like the dazzling shots of far-off galaxies taken by the Hubble Space Telescope, the original black-and-white micrographs are often colorized to highlight specific features of the virus. Three-dimensional computer graphics provide even finer details of HIV's internal and external architecture. I know that these images have been invaluable to scientists in their growing understanding of HIV and in crafting new models for fighting it, but, to me, the greater the complexity of the virus, the bleaker the chances seem of surviving it. The discovery of a cure feels farther off in my lifetime and unlikely in Steve's.

During rough patches, Steve admits that his life seems to creep forward in three-month intervals, alluding not to the turn of seasons but to the stretches between his getting blood work, the panoply of tests that measure the virus's activity and how well his immune system and organs are holding up. The findings provide an assessment of his current drug regimen and help determine the course for the next twelve weeks. Seeing his doctor for the results is always anxiety producing.

Thinking back fourteen years, I don't remember Steve ever having a slim patient file, although at some distant point in our shared past that must've been the case. Now it's a thick sheaf that's plopped onto the desktop at the start of each appointment. Once the pleasantries are over, Dr. Hassler opens the file and the three of us huddle over the latest labs, a three-page printout of more than fifty separate tests. The results run down the center of each page in one of two columns: WITHIN RANGE, under which most of Steve's liver and kidney function results, for example, are listed; and OUTSIDE OF RANGE, where the grimmer numbers, T helper percentages, white cell counts, and the like, are clustered. To make the bad numbers easy to spot, this column is shaded a pale red, a stripe top to bottom. The most recent pages are bound to a thick pile of Steve's past results, and a quick fanning of the stack creates a crude animation of a red ribbon, many years long.

ACROSS THE BAY BRIDGE AND THIRTY MILES FROM MY HOME, I STEP from my car and approach IDL, Immunodiagnostic Laboratories. The broad one-story building, here on the outskirts of San Leandro, is situated in a secluded industrial complex. The building proper is faced entirely in black reflective glass, making it impossible for me to glimpse any activity within. What's more, while a sign confirms that I am definitely at the correct address, I can't find a front entrance or even the appearance of a door. *How apropos,* I think; *the lab where Steve's blood is tested is testing me.* Stumped about how to get inside, I stare at the building. All I see is myself, looking back.

I walk around to what appears to be the rear of the building and, at last, locate a doorbell.

A head shoots through a cracked-open door: "You got a delivery?"

"No, an appointment," I reply.

"Well, this is shipping." Nevertheless, the young man agrees to take me through the building to the front office, where I'm left to wait in a blah-colored reception area that is conspicuously missing a receptionist. The phone rings over and over. I know she exists because I spoke with her yesterday.

"IDL," the woman had answered.

I'd hastily dialed the number stamped atop the lab request form Steve takes to the phlebotomist and hadn't quite formulated the nature of my request, nor to whom I wished to speak. The Chief Blood Tester? Blood Docent? "I'm wondering," I stammered, "could I come out and just take a look around?"

"A look around what?" she returned pleasantly.

"Well, at the lab. Get a sense of the process involved with testing blood. Maybe, if it's not too much trouble, take a tour of the facility."

Long pause. "A tour?" I pictured her scanning the room, thinking, *Heavens, what guidebook does he have?* "Um, well, we don't *give* tours. Are you sure you have the right phone number?"

Oh, yeah, yeah, yeah, I explained, I didn't need any blood tests myself but was interested in seeing how they're done. The more I spoke, the odder it must've sounded. As if she'd finally caught on, the receptionist said, "Oh, are you with the FDA or something?"

Before I could deny anything, she'd transferred me to IDL's medical director, the head honcho, Edward Winger. He, thankfully, understood my desire to see what happens to blood in the in-between, after it's drawn but before the results are sent on.

Sure, he could show me how it all works. "How about ten-thirty Friday?"

Before hanging up, I thought, *Oh, go ahead, just ask:* "My partner just had his blood drawn," I began, "and I assume you'll have it by the time we meet. Would it be naïve of me to think that I could actually see his blood being tested?"

His burst of laughter supplied a thorough answer, but, lest I had any doubts, Dr. Winger added, "Yes, it would be naïve of you. We don't track the blood by a person's name."

"Oh, for confidentiality purposes, you mean," I said. "That makes sense."

"But also," Dr. Winger added, "almost all of the testing is done late at night."

Night? So the workers come out to count the blood only after sundown. How vampiric. Well, no wonder parking was so easy.

Dr. Winger enters the reception area. He's a tall, thin man in his fifties. He has silver-blue eyes behind delicate wire-rimmed frames. Shaking his hand, I find it cool and powdery-dry, as if he has just pulled off a latex glove.

Without further delay, Dr. Winger begins the walk-and-talk, ushering me into the laboratory he founded in 1982. I can see right away that the word *laboratory* doesn't quite fit, associated as it is in my head with beakers, bottles, and burners. Immuno-diagnostic's lab is a facility about the size of a basketball court—brightly lit, with white walls and shiny floors. But chilly. Now I understand why Dr. Winger wears a heavy flannel shirt on this Indian summer day. I spot a total of three people in work areas scattered about the floor. In lieu of introducing me to his staff, however, over the next hour Dr. Winger will introduce me to his machines.

The first pair handle what is called viral load testing, which provides a measurement of the amount of HIV in a person's blood. A decade ago, the best test of this kind was the p24 antigen, which only gave a *Yes* or *No* answer to the question, *Is the virus actively replicating?* It worked by searching the blood for a discarded part of HIV, a method akin to determining if a McDonald's burger has been eaten by rummaging for the tossed wrapper. By contrast, today's viral load tests zero in on the Big Mac itself, the genetic material in HIV. They quantify how virulent your virus is as well as whether or not the pills you're taking are having an effect. The two machines before me aren't large or imposing, but their power to change a person's life is enormous. How the tests work, though, is complicated, and I pay a price for my momentary lapse in attention. Dr. Winger is in the midst of describing the most sensitive of the three types of viral load tests, the Q-PCR:

"...so we have a single-stranded molecule and another single-stranded molecule here"—Dr. Winger is now also drawing—"and what happens is, we end up having only this region here being copied, and then, well, then we get a double-stranded molecule."

He makes a new addition to his notepad. "You see this?"

Yes, I see what looks like a drawing of venetian blinds—strips of flattened DNA, I gather, nodding. But Dr. Winger has already moved on. He draws two graphs that look like sales curves.

"With each cycle, we get a doubling of just this region here and it amplifies geometrically . . ."

Dr. Winger's verbal momentum is gaining speed, yet I am utterly lost and feel only a little regret at pulling out his power cord. I ask for the layman's version and he obliges, although, at first, it is still more of the Dr.-Layman-Ph.D. variety. But finally, he breaks it down this way: They take a sample of blood—less than half a teaspoon—then remove a single fragment of DNA from an HIV particle and clone, or "amplify," it. Using a mathematical formula, they then calculate the number of viral particles, or "copies," originally present in the sample. This number is an accurate fraction of the total amount in the bloodstream. *Okay, that makes sense,* I think, but Dr. Winger cannot resist a big textbook finish: "There's a rule of thumb that the number of cycles required is inversely related to the log of the starting copy number."

What's not lost on me is the impact these results will have in the doctor-patient sit-down. There, it's not a number you hope to see but a word. When fewer than fifty copies are found in a patient's blood, the Q-PCR test finding is labeled "undetectable." Fifty copies may sound like a lot, but this is actually an infinitesimal amount of HIV. The take-home message is, if your virus is undetectable, your drug cocktail's working and viral activity is at a virtual standstill. Though its diagnostic meaning is unambiguous, the casual use of the word has caused problems. When doctors announced in 1997, for example, that Magic Johnson's virus was undetectable, many fans took this to mean that the former Laker no longer had HIV. It didn't help that his wife, Cookie, declared in an *Ebony* magazine interview that Magic had been "cured." His doctors "think it's the medicine," she's quoted as saying. "We claim it in the name of Jesus." But no miracle had occurred. In fact, after Johnson later neglected to take his meds during a long vacation, his viral load shot back up to detectable levels. My partner Steve has put his own spin on this semantic confusion: *Undetectable* is a lot like the Invisible Woman from the Fantastic Four—just because you don't see her doesn't mean she's not there.

Dr. Winger explains that, of course, PCR testing has other applications outside of HIV care. It serves an essential role in forensics science, for instance, by isolating the DNA "fingerprint" of blood or tissue evidence found at crime scenes. *And the killer is . . . !*

At this point in the tour, I'm realizing how loudly he and I have to talk to be heard over the racket made by these two machines. (What must this place sound like in the middle of the night, when all the machinery is in high gear?) We stand before the apparatus that is used to isolate the DNA molecule. Though it's not much larger than a toaster oven, it sounds like a dryer filled with tennis shoes. The thumping, Dr. Winger explains, is made by a piston that pushes cell particles through an interior tube at a pressurized weight of more than three thousand pounds per square inch. I mouth *wow* back to him.

We now move on to other noises. Loudest of all are two liquid nitrogen tanks. These are powered by individual generators that are doing a good imitation of cement trucks. The tanks, which resemble a large pair of bongos, are where tissue and cell cultures are preserved. "They're cold as hell," Dr. Winger specifies. "Minus 195 degrees Celsius." As he unlatches one of the lids, fog-like vapor overflows. "Put your hand down in there," he urges me, flashing a triangle-shaped smile. "But don't touch the sides!" I'm reluctant—I've seen *The Empire Strikes Back* too many times not to flash on Han Solo being frozen in carbonite— but I summon the wherewithal to dip my index finger in partway. "Very, very cold," Dr. Winger chirps. "Exceptionally cold."

Continuing on in the doctor's wake, I realize that I haven't seen any blood anywhere in the lab, not a drop of red in the sterile sea of white and black and bland equipment. I'd expected to see rows and racks and stacks of vials. But, as in the body, the blood at IDL is just under the surface. It's concealed within machines. It's stored behind refrigerator doors. For certain tests, it's kept in incubators, body-temperature warm. I know that there are five vials of Steve here, someplace on the premises.

Every specimen that enters this facility is stripped of its identity, Dr. Winger tells me. Each vial is bar-coded, and its every move through the lab is monitored by computer. This makes for the easy and accurate assembling of test results from many machines. The computer does not make mistakes, he states evenly. It strikes me that the whole process is, as much as possible, devoid of human touch and emotion, but also of human error and carelessness. Unbidden comes the memory of that letter Steve received from his previous lab with the news of the phlebotomist who'd reused needles. But I'm comforted by all that surrounds me. His blood's in good hands here, during this part of its journey, just as he himself is with his regular IDL phlebotomist, Rosemary.

Pausing for a moment in the center of the lab, Dr. Winger quickly points out some of the noteworthy machines around us: "This is an ELISA reader, Western blots here, blood chemistries over there. Immunochemistry stuff. Urinalysis. Blood coagulation panels. Over there is the DNA synthesizing machine, which is synthesizing as we speak." Across the room, that's an ultracentrifuge, a device for spinning plasma at superhigh speeds—"It's forty thousand times gravity in that thing"—a process that separates the component parts of cells.

Amid all this expensive high-tech equipment, I spot something familiar. "That looks almost like a microwave," I say.

He grins a *you-got-it*. "There's nothing better than a microwave for making basic heat-dried stains," he admits. "It's the only thing here under a hundred dollars." This last bit he adds with an air of amusement. They'd just had to send back a quarter-million-dollar piece of equipment that had turned out to be a real lemon. Go figure.

As I follow Dr. Winger toward the last stop on my tour—the T cell tabulator—my mind wanders backward. T cell counts, unlike, say, the newer viral load tests, have been the through-line of Steve's long life with HIV, albeit a through-line with peaks and plummets. Of the various T cells counted—helpers, killers, and suppressors—the helper T's are the most important indicator of how your immune system's doing in fending off the virus. In a healthy person, a normal helper count—often simply called T cells, for short—could be as high as eighteen hundred per cubic millimeter of blood; in a person with advanced HIV disease, it could be eighteen, or zero. Falling below two hundred is the criterion for an AIDS diagnosis. This truck hit Steve in the summer of 1994. Following that, he had to get T cell counts every four weeks as his immune system continued to deteriorate. Watching those numbers descend was a helpless feeling, since Steve had already done every antiviral available and the next wave of meds, the protease inhibitors, was still a year off. It was like he was stranded in the desert and could only watch as his water supply fell.

In the early years of the AIDS epidemic, T cells—as well as all blood cells, for that matter—were counted by hand. In my head I pictured row upon row of white-coated lab techs, all hunched over microscopes, quietly tallying cells with calculators, and all, in a curious casting choice, middle-aged women. The row of ladies who tallied T cells looked more beleaguered than the rest, I imagined. I actually worried about them, faced day after day with the blood of the very ill. I hoped they gave out a private hoot when a robust sample came through. In some parts of the world, these kinds of counts are still done manually. During a recent tour of an AIDS clinic in Rwanda, a friend told me, he watched as a woman laboriously counted blood cells, an eye to a microscope lens, a finger on a simple clicker.

Dr. Winger and I stand before the Flow Cytometer, the state-of-the-art cell tabulator, a machine that, to me, would not look out of place at a Kinko's. He introduces Mark, the technician who operates it, but then backtracks a bit to remind me of a basic fact of hematology: White cells look a lot alike. While it's easy to tell the difference in a blood smear between, say, a red and a white cell, the distinctions among the types and subtypes of lymphocytes are subtle. "You can't tell a helper T from a suppressor T cell under a conventional microscope," he explains. But there's a way around this. By introducing into the blood sample what's called a monoclonal antibody, the specific white cell you're trying to count will be "tagged." Next, a dye is added that stains the tagged cells.

"How very Paul-Ehrlichean," I comment.

"Exactly. It was his idea to couple antibodies to dyes and use them to identify cells."

"But today this is all done by computer."

Dr. Winger nods. The dyes used are fluorescent, which makes them recognizable by laser. He next points to a rectangular black contraption, the contents of which aren't visible. "We put a test tube of blood in a carousel down there, and Mark here tells the computer we want to 'interrogate' certain stained cells. So, for example, it allows us to look at T helper cells only."

"You say *look at them,* but you're never looking at the cells directly."

Well, no, he concedes, but the computer is. "Every single cell passes by a sensor head that inspects it." At the same time, the flow of cells is shown on a computer screen. Sure enough, Dr. Winger points to a monitor where a meteor shower of gold pixels is shooting across a black field, left to right. These are T cells. I have no reason to believe they're some of Steve's, but then again, who knows? Either way, I find myself transfixed, rooting for a high count. I wait until I'm sure several hundred have flown by. Now it's safe to move on.

Blood Criminal

THE CRIMINAL TRIAL OF THE SMITHKLINE BEECHAM phlebotomist accused of deliberately reusing blood-draw needles was scheduled to commence in mid-August 2001, more than two years after Steve and thousands of other patients had first been notified of this woman's dangerous actions. How often she'd reused the butterfly needles and with which patients remained unknown or, at least, unreported. Either way, the math didn't look good. She had been employed by the lab off and on over six years, during which she'd had contact with up to twelve thousand people. (In a sickening coincidence, the last of the eighteen times Steve used a SmithKline lab resulted in the blood work that gave him his AIDS diagnosis.) He and I had never seen a photo or news footage of the "renegade phlebotomist," as she was called in some early media reports, and we didn't learn her name until a May 2001 newspaper story provided details of her upcoming jury trial. Elaine Giorgi faced six felony charges, including assault with a deadly weapon—dirty needles. Though she'd worked at SmithKline labs throughout the Bay Area, the charges had been filed in Santa Clara County. The trial, in which she'd be represented by court-appointed attorney Brian Matthews, was to take place at the San Jose Hall of Justice.

I wasn't surprised to learn that the first session served as an opportunity for her lawyer to request a delay, which was granted. But I never expected that, over the next year, her trial would be postponed ten times due to assorted legal matters. I was anxious to get a good look at Elaine Giorgi and to hear how she justified what she herself had admitted to doing "occasionally." And then, in July 2002, the prosecutor dropped the most serious assault charge and Giorgi pleaded guilty to separate felony violations of illegally disposing of medical waste. There would be no jury trial after all, only a sentencing. She faced a slew of fines and a maximum of five years in state prison.

The picture I had in my mind of what a blood criminal looked like ran a continuum, from the sublime to the horrific. During the summer of the bloody glove and the O. J. trial, a trio of Italian bank robbers stole a couple of watts of the media spotlight after helping themselves to five-figure sums from more than ten Turin banks. Up through their capture, their brazen acts won them cheers, especially from people with HIV, for these Bonnie-less Clydes had AIDS and a legal loophole on their side. A 1993 "compassionate release" law in Italy prevented the terminally ill from serving jail time. In those days before effective drug cocktails, the three's spree demonstrated a fearlessness that, here in San Francisco, raised a few spirits and felt downright therapeutic. *Silenzio = Morte!*

No smiles surfaced two years later for Nushawn Williams. This nineteen-year-old, arrested in 1997 in New York State, became the face of criminal HIV transmission when he was accused of deliberately infecting thirteen young women, including an eighth-grader, through unprotected sex. Each new revelation added to the horror. He traded drugs for sex. He kept records of his exploits. He may've exposed nearly fifty individuals. He said he didn't believe the social worker who'd told him the previous year that he was HIV positive. Williams later pleaded guilty to four sex-related felonies, including statutory rape and reckless endangerment. He was sentenced to four to twelve years in state prison, where he's currently doing time. His parole was denied in 2001, and again in 2003.

Now joining this rogues' gallery was Elaine Giorgi.

Steve and I woke early to make the drive to San Jose for the 8:45 A.M. court date. Mile after mile down Highway 101, traffic to our right was a clogged artery as we zipped by in the diamond lane. We arrived with a couple of minutes to spare, only to find the doors of the third-floor courtroom locked, which caused us some distress but seemed not to worry any of the people seated in the hallway. We sat down on the banquette that stretched the length of the windows. After all the trial delays, it now struck me as silly to have thought the sentencing would start on time.

I tried to figure out who all these folks were—the solitary older gentleman dressed in khakis and a T-shirt; the two younger men in navy-blue suits down to the right, lawyers presumably; the man and woman across from us, knotted in conversation. The brittle, waiting-room atmosphere changed in a snap when two more besuited lawyers strode from a courtroom down the hall—both wearing the grin of a happy verdict—and paused to talk to the couple opposite us.

Steve leaned in close. "Do you think that's her?" he whispered. Beside the doughy man in his midthirties sat an unremarkable silver-haired woman, fifty-something, in a black pantsuit.

"No, I don't think so." Over the yawn of months, I'd constructed a mental image of the phlebotomist, or at least a rough outline. In my head she was tall and fleshy and robust. Her sheer physicality accounted, in part, for her ability to deceive and frighten so many people, to inflict such damage. By contrast, the hunched woman seated a few yards away looked tiny and frail. Her shoulder blades jutted from her suit jacket, as if she'd forgotten to remove the clothes hanger when she'd dressed.

"It *is* her," Steve said, his voice soft but insistent. "She's with her attorney. Matthews, right? One of those lawyers who just walked up said his name. That's got to be her."

"I . . . I think you're right."

Steve's face said pure relief and I knew why: He did not recognize her. For three years, he'd been worrying about the two phlebotomists who'd regularly drawn his blood at the San Francisco lab. Though he couldn't remember their names, he could picture their faces. This lady was neither of them. Which meant that his blood had, in all likelihood, never caused anyone harm. His *phew* was a lovely sound.

Just then, a deputy emerged from the courtroom and propped open the door. Joining the queue, we followed Elaine Giorgi, her attorney, and a handful of others into the small courtroom. I sidled into the row behind Giorgi's. As I reached down to lower the seat, I studied her appearance. The first thing I noticed close up: The silvery hair was a wig. She tugged at it and the whole helmet shifted. Either it had stretched or she had shrunken. With her wire-rimmed glasses, lined face, and that hair, she resembled one of the Golden Girls, Estelle Getty's character.

We'd hardly settled into our seats when the deputy approached and told a few of us that the first portion of the hearing would be closed to the public. As quickly as we'd entered, we were headed back out into the hallway. The deputy was very pleasant and said he'd notify us when we could return. The courtroom door shut. I was surprised by how few people had been inside—twelve at most, including the officers of the court. I'd expected many more, given the media attention that had swirled around this case three years earlier. Where were all the reporters? The TV cameras? In fact, we now waited with just one other person, the older gentleman I'd spotted before. "You must be here for the Elaine Giorgi case," I said.

"Yes," he answered, "I'm one of her victims."

Jerry Orcoff was a big man, about six feet tall, with a bristly white beard and glasses. He looked like the kind of guy who'd give you a good deal at a flea market. As strangers in the same boat, we naturally began to share our stories. Four years ago, Jerry said, he was diagnosed with hepatitis C. At the time, he'd had no idea how he could have contracted this viral infection that's most common among IV drug users—which he'd certainly never been. Then came the letter from SmithKline Beecham. Jerry realized that the dates matched up—he'd first fallen sick with the characteristic flu-like symptoms shortly after he'd had his blood drawn, one time only, for some routine medical tests. He'd gone to the Palo Alto branch of the lab. He was certain Elaine Giorgi had been his phlebotomist.

Jerry was a retired mechanical engineer, seventy years old, married, and the father of two grown children. Sure, he'd expected to slow down a bit as he grew older, but he never—Jerry couldn't finish the sentence. He just gave me a look like his dog had died, then shook his head.

Viral hepatitis, as I knew, has six different strains, all represented with letters. A, B, and C are the most prevalent. Hepatitis A is spread through water or food contaminated with fecal matter (every school year seems to come with a news story about kids who've been exposed through improperly washed fruit; frozen strawberries are a common culprit). Hepatitis B, like HIV, is most often transmitted through unprotected sex. By contrast, the hepatitis C virus—HCV for short—is exclusively blood-borne and spread primarily through shared dirty needles; less frequently through accidental needle sticks in health care settings and from mother to newborn during childbirth; and, in a small number of well-publicized cases, through shared or unsterilized tattooing equipment. Whatever its initial cause or strain, viral hepatitis can destroy the liver's ability to perform life-preserving functions, including filtering toxins from the bloodstream and converting blood sugar into usable energy. The most visible sign of advanced disease is jaundice. This yellowing of the skin and eyes indicates that the liver is failing to clear the blood of what are called bile pigments, the yellow-colored byproducts of dead red cells. Vaccines now exist to prevent hepatitis A and B, but not C.

HCV, the most common chronic viral infection in the United States, has been labeled "the silent epidemic." It's difficult to treat, has no definite cure, and in most cases lingers for years without expressing symptoms. In fact, the virus's progression can be so slow that an infected person is more likely to die of other causes. But in 15 to 20 percent of cases it quickly brings on cirrhosis, liver cancer, or related illnesses. HCV is the leading cause of liver transplants in the United States.

Jerry, who's still in a relatively early stage of chronic HCV, is plagued by fatigue, he told us, but not by worse health problems—knock on wood. His disease, which began with a single blood draw, has led to many draws since, he explained, as his progress is monitored through tests of liver enzymes and other markers. "I'm fighting a dragon," he said, and I first thought he'd chosen an apt metaphor for illness. Then I realized he meant his lawsuit. Like other former patients of Elaine Giorgi, Jerry had filed a civil suit against her former employer, SmithKline Beecham, which is part of one of the largest pharmaceutical companies in the world. A few weeks back they'd offered him a settlement of a few thousand dollars, which he refused. He told us he was hoping his case would go to trial within the year, though he wasn't betting on it. Dragons are so good at dragging their feet. Regardless, today Jerry Orcoff's stake in this whole awful mess was not financial but emotional. Today he wanted to see Elaine Giorgi brought to justice. He would have to wait for that, however, as would Steve and I. Right then, the deputy reemerged and explained that, due to some paperwork missing from a probation report, her sentencing had been postponed until August 15.

With a wave, Jerry said, "See you next month," and strolled off.

THE SANTA CLARA COUNTY PROSECUTOR IN CHARGE OF THE CASE, Dale Sanderson, explained to me that Elaine Giorgi's bizarre behavior—his words, not mine—went way beyond her bone-headed reuse of needles—my words, not his. For instance, he elaborated, a co-worker had caught her deliberately putting the wrong patient's name on a vial of blood. Evidently Giorgi hadn't taken enough blood from someone who'd already left the lab, which isn't an uncommon mistake. It happens. But rather than calling the patient back in for another draw, Giorgi made up the difference using someone else's blood. Although I was speaking with Sanderson on the phone, clearly he heard my jaw hit the floor. He was just as aghast. "I mean, can you imagine?" he exclaimed. "Can you imagine going in for your blood results and being told you have a disease that's not even in your blood?" Or, say, your wife going out partying with friends because she'd been told she wasn't pregnant when, come to find out, she was?

Giorgi then tried to cover up her mistake by changing the blood work requisition form that the second patient had brought in, so she could take the extra vial. Under the law, these actions, along with other violations she'd committed, were misdemeanors. But Deputy District Attorney Sanderson wanted a felony conviction and its stiffer penalty. Looking back three years to when he'd first received her case, he recalled, "I figured it would be very easy to show that someone who reused a needle puts the entire world at risk." But that's not how it turned out. Unable to find a corresponding law on the books, Sanderson thought back to a murder case he'd prosecuted in the late 1980s, "the first pit bull killing case," in which the dog's owner was charged with using a violent animal to commit assault. Sanderson believed this case showed a promising parallel to Giorgi's reuse of potentially deadly needles. In addition, he "dusted off" a statute in the California Health and Safety Code regarding the illegal treatment or disposal of medical waste. Although no health care provider in California, to Sanderson's knowledge, had ever been prosecuted under this statute, he felt sure he could use it to argue Giorgi's culpability on multiple felony counts— that in reusing needles she was unlawfully "treating" biohazardous waste.

Sanderson's strategy withstood the numerous legal hurdles

of the preliminary hearings, he explained to me, but, just days before Giorgi's trial was to begin, the California Supreme Court pulled a significant patch of rug out from under his feet. In its ruling on an unrelated case, the court narrowed the legal definition of the word *likely* (in the charge of "assault with a deadly weapon likely to cause great bodily injury"). This narrowing made it unlikely that Sanderson would get a conviction on Giorgi's assault charges. In exchange for dropping these, she agreed to plead guilty to the remaining "less egregious" felonies, as Sanderson called them, as well as a single misdemeanor.

The woman who awaited sentencing on the afternoon of August 15 looked different from the one Steve and I had seen in court four weeks earlier. Elaine Giorgi's silver wig was gone, revealing dyed rusty hair in a ratty cut. If possible, she appeared even thinner and more exhausted, as if she were also suffering the full weight of Jerry Orcoff's fatigue. While the courtroom of superior court judge Hugh F. Mullin III was electric with last-minute activity as the legal teams prepared to begin, she sat frozen.

"So what's with the hair?" I said, sotto voce, sitting in between Steve and Jerry a few rows behind and to the left of Giorgi. Among the three of us, two theories quickly emerged: Wiglessness was either a ploy on her attorney's part to make her look as pitiful as possible before the judge, or it merely served a practical consideration—if she were taken straight off to jail, her personal possessions, silver helmet included, would be confiscated, and perhaps in her thinking the day would bring indignities enough. Steve pointed out that her tailored pantsuit was gone, too. The ex-phlebotomist, now a felon, wore informal slacks and a sweater, clothes I doubted she'd have given a second thought to bundling up and using as a pillow.

At Judge Mullin's entrance, the courtroom clockwork instantly wound down. With his small build, full ruddy cheeks, and clipped mustache, he looked like the tycoon from the game Monopoly, except for his expression, which was, already, one of profound irritation. Giorgi's attorney, Brian Matthews, wasted no time and called to the stand forensic psychologist Rahn Minigawa. Minigawa had spent seven hours interviewing Giorgi and administering personality assessment tests. From TV courtroom dramas, I suppose, I was primed to think that a photogenic authority such as he would provide a lucid, penetrating explanation of the deep-rooted cause of Giorgi's behavior. But in this instance, I experienced no eureka-like *aha!* as Minigawa ticked off the phlebotomist's laundry list of personal problems, from childhood physical abuse to a more recent history of depression, panic attacks, and compulsiveness. A recovering alcoholic, the fifty-five-year-old Giorgi had had other legal problems stemming from two DUIs as well as a host of financial troubles, he testified. What's more, "She's also menopausal." It was the doctor's belief that the debilitating side effects of menopause, such as the insomnia and mood swings, together with the aforementioned problems—*Okay, here comes his conclusion,* I thought at the time, *the distillation of seven hours of psychological analysis, the meat of her defense*—"All affected her ability to make good decisions."

Throughout all this, Jerry reacted with a steady emission of sighs and *harrumphs,* which I believe is the straight man's way of saying, *Oh please.* Steve turned his notepad to me and pointed to where he'd doodled a bewigged stick figure wearing a T-shirt that read MENOPAUSE MADE ME DO IT. But I quickly returned to watching Judge Mullin, who, chin propped on fist, continually re-aimed his gaze at Giorgi.

Next up was Dale Sanderson, whose every word and manner

in his cross of Minigawa said, *Gimme a break.* The prosecutor did not hide his disdain for this paid defense witness, this "expert." Pressed by Sanderson to dispense with the DSM jargon and to provide the court with genuine insight into why she'd done what she'd done, the doctor hesitated, obviously uncomfortable, then replied, "Well . . . she said she didn't have a clear recollection of her actions . . . she said she was stupid."

Sanderson, who throughout the proceedings was never granted the opportunity to interview Giorgi, wanted to make sure he had heard Minigawa correctly. Okay, he recapped: You administered two personality tests. You spent several hours with her. You're an expert who's testified in more than fifty trials. And the best assessment you can provide the court of her motivation is that she said she was "stupid." With an expression I found vicariously satisfying, Sanderson looked at Minigawa as if the doctor were something he'd poked with a pencil from the bottom of his shoe.

Finished with the psychologist, the prosecutor laid out his own theory of what had occurred. Elaine Giorgi's actions were calculated, he contended, and were motivated by the desire "to curry favor with SmithKline Beecham and make patients happy." Giorgi, who'd once been fired from the company for her inefficiency, sought to make the most of this second chance. The problem was, she wasn't a very good phlebotomist. She had to perform thirty to fifty blood draws per day and, despite her training and experience, had a hard time using the standard needles. She found the small, light butterfly needles easier to use. Patients also found them less painful; hence fewer patients made complaints. Just one drawback, though: Butterflies were expensive—eighty cents apiece as opposed to five cents for a standard needle—and they were intended for use with only a small number of patients, mainly pediatric and geriatric. Giorgi reused the pricier butterflies because, if she ordered them in mass quantities, her bosses would notice.

There was my *aha*.

Sanderson said in closing, "It is inexcusable that she'd value a human life at seventy-five cents."

Matthews spoke next—a candle to Sanderson's fireworks—and meekly summed up with, "I don't think society needs to be protected from her."

Finally, Elaine Giorgi read a prewritten statement. Her quavering, peeping voice obscured each syllable before it reached my ears, but the judge nodded at her when she'd finished.

Up to this point Judge Mullin had said little beyond requesting the occasional clarification from one of the attorneys. Now he took aim at Giorgi and said in a voice of seven thunders, "What you did was as dangerous as holding a loaded gun to

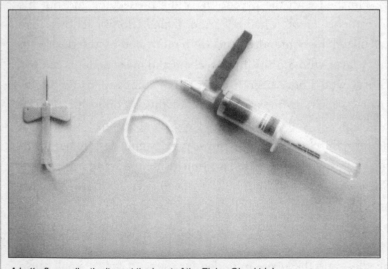

A butterfly needle, the item at the heart of the Elaine Giorgi trial

your victims' heads." He then paused, whether to wring the disgust from his voice or to let it build up, I couldn't tell. And those patients, he resumed, "were as vulnerable as you can get." She was lucky beyond words that no one had died of AIDS or another fatal disease due to her actions, a comment that was powerful but, it struck me, inaccurate—HIV rarely progresses that quickly. Still, obviously, the judge wasn't out to educate the crowd. Under this barrage, I don't know how Giorgi was able to remain standing.

Judge Mullin then broke from addressing the defendant and spoke more broadly to the court: Had someone died, he declared, she'd be facing a long stretch in state prison. "In this case, prison would—" He stopped and shuffled through some papers. "—do absolutely no good. Except as punishment."

Wha—? I'd been right there with him up to that point, but . . . Was she going to go free?

"But I *am* going to punish her. To a year in county jail, plus four years probation."

For the next couple of minutes, while the various fines that Giorgi would have to pay were entered into the record and the details of her future parole were clarified, my brain kind of zoned out. I found myself with that surreal sensation one has after enduring a long flight—your plane's landed but it's still taxiing—that contradictory feeling that you've reached your destination though you're not quite there yet. Around us people started standing and we fell in line, following the folks in front of us toward the exit.

I took a deep breath of sidewalk air and realized I'd arrived at a moment I'd never thought beyond. Jerry, who'd steered us out of the Hall of Justice, seemed to have already placed the day into a healthy perspective. Yes, he agreed with Steve, he was disappointed by how short a sentence she received but, in the long run, he was glad that she would never be permitted to draw another person's blood.

Behind us, on the opposite side of the courtyard, reporters had congregated. "Are you a victim? Will you speak for the camera?" they called out to passersby in English and Spanish. *"¿Es usted una víctima?"*

At the same time, Giorgi emerged, clutching the arm of her attorney. For reasons unclear to me, she was being allowed to walk herself to jail, which was located half a block away. The reporters swarmed but she didn't say a word. Her attorney held up a hand, *No comment,* and the two kept moving. With that, Jerry headed to his car. Steve and I stood and watched as Elaine Giorgi climbed the last few steps that led to the Santa Clara County Jail.

NINE

Exsanguinate

Blood makes noise
It's a ringing in my ear
And I can't really hear you
In the thickening of fear

Blood makes noise

—SUZANNE VEGA, 1992

FOR A PERIOD OF SEVERAL HUNDRED YEARS UP THROUGH the nineteenth century, the earthly punishment meted out to British criminals sentenced to death did not end with execution. Worse than imagining the tightening of the noose or the plunge of the guillotine's blade, according to writings of the time, was a felon's fear of ending up on the wrong side of an anatomist's knife. The idea of your body being sliced apart piece by piece—regardless that this was done to instruct medical students or in the name of science—could tap into every private horror, whether humiliation or desecration or something grislier. For this, the condemned had England's Henry VIII to thank. In 1542, by royal decree, the Guild of Barbers and Surgeons—the bloodletting specialists who sidelined in haircuts and minor surgeries—was granted a maximum of four executed "malefactors" per annum for use in public dissections. This was the only legal source of cadavers. By no miracle of accounting did four per year come close to meeting demand, however, and the shortage led to a thriving black market in stolen bodies.

In 1752 the king's law was amended to allow a judge to

The Reward of Cruelty
by William Hogarth, 1751

send any executed convict's body to the Surgeons' Hall. A felon had just cause to worry. An engraving titled *The Reward of Cruelty* (1751), by the British artist William Hogarth, depicts a dissection-in-progress at Surgeons' Hall. The naked body of a freshly executed murderer is splayed in a crowded auditorium of onlookers, and the lead anatomist directs the activity with a long stick: *Cut here and gouge there, if you please.* One surgeon pries loose an eyeball, another slices open the foot, while a third man seems to have slid his entire hand up into the deceased's chest cavity, perhaps reaching for the heart. A final man kneels to the side collecting in a bucket the long sausage of the intestines. While Hogarth's engraving is a work of satire—the noose is still affixed to the felon's neck, for instance, and a small dog is about to make off with what looks like the liver—it nevertheless captures the graphic nature of the butchering.

More gruesome still were twin dissections performed the following century, as recounted by medical historian Gustav Eckstein in his book *The Body Has a Head* (1970). Eckstein's tale is thin on personal details but rich in methodology. Two criminals, sentenced to decapitation, would be used to answer once and for all the nagging question, *How much blood does the human body contain?* Of course, many times throughout history best guesses had been made, but this latest effort would be as exacting as humanly possible. First, each man had blood drawn—a predetermined amount, which was diluted precisely one hundred times. These samples were set aside to serve later as a color standard. With no further ceremony, the two men were relieved of their heads, and all spillage was collected. The heads and trunks were drained, then squeezed. Once no more color would bleed, the bodies were carved into smaller and smaller pieces, eventually down to human chum, and washed, soaked, and wrung. All excess water added to the process was tallied and saved. Finally, in a process that to me seems fraught with the potential for error, the total liquid remains of each criminal were color-compared to his original sample, dilutions were made until they matched, mathematics were applied, and both weight and volume proportions were calculated. The results, consistent in both men, were presumed representative of all people—not inaccurately, as it turned out. The exsanguinations "proved our blood to be one-thirteenth of us," to borrow Gustav Eckstein's summation. This is roughly 7.5 percent of our total body weight. Likewise, every thirty pounds of us has about a quart of blood. For the average

150-pound man such as myself, that's 11.25 pounds of blood in circulation, or, echoing Eckstein, "Five quarts go the round-and-round."

Now, trading horror for horror, the coolness of science for the seduction of literature: Set in the same century, this next story revolves around the same unsavory deed—the taking of blood—yet to an altogether different end and employing a more sensual methodology. The basic plot should be familiar, even if you've yet to read the tale in its original form. Within a handful of pages, our protagonist stands in the gloom of a desolate night, in a foreign land, on the doorstep of an enormous stone castle. He finds no bell or knocker and is unsure how to signal his arrival after the arduous journey from London. Just then, a noise from within: rattling chains followed by the clanking of massive bolts, a key thrust into the lock. At last the door swings open. Centered in the entryway stands a tall older gentleman dressed in black, clean-shaven but for a long white mustache. His pale, pale skin picks up none of the warmth cast by the flickering lamp he carries. "Welcome to my house!" he says in peculiarly inflected English. "Enter freely and of your own will!"

The weary traveler shakes an ice-cold hand, and the elder man makes it official: "I am Dracula."

If you, too, are still puzzling over that *long white mustache*, I'm right there with you. The Dracula in Bram Stoker's *Dracula*, the template for bloodsucking horror since its publication in 1897, does not resemble Bela Lugosi; nor does sunlight destroy him. The story of the Transylvanian vampire has been retold and reimagined in such varied contexts—from early Hollywood flicks to adult films, a soap opera to video games, and a breakfast cereal to the number-obsessed Count von Count from *Sesame Street*—that it's fascinating to discover elements in the original source that, more than a century later, feel new. How creepy, for instance, is Dracula's lizard-like way of scaling walls. And how clever is his strategy of concealing fifty coffins throughout the greater London area so that, after prowling, he has a wide variety of places to rest during the daytime. And, oh yes, the way the heroes use Communion wafers to render these coffins unsleepable. At the same time, it's great to find all those familiar trappings of vampire lore: the mirrors that don't reflect, the fangs, the turning into a bat, the garlic, the stake through the heart. And sure enough, the greedy mouthfuls of blood. Save for the Bible, no other work in the English language has had, I'd wager, a stronger impact on how people of the modern Western world think and feel about blood. Blood as dangerous and profane as opposed to sacred and profound.

Stoker's novel, originally titled *The Un-Dead,* a term the Dubliner coined, is more ambitious than I had remembered, technically as well as psychologically. But *Dracula* is also very much a dusty product of its era. Abraham (Bram) Stoker (1847–1912) wrote a conventional Gothic novel, the type of romantic fiction that first appeared in England in the mid-eighteenth century, a forerunner of the bodice ripper and the modern mystery novel. True to the form, *Dracula* features a damsel in distress (two, actually); a good guy (in this case, a quintet of good guys); and a tall, dark villain, although here, obviously, Stoker created a new standard of darkness. As was typical of Gothic fiction, the action takes place in ominous locations, shadowy and perilous, the most archetypal of which is the count's home, Castle Dracula.

Stoker wrote the novel during a seven-year period that neatly falls between two major advances in the understanding of blood: the identification in the 1880s of platelets, the circulating blood cells that aid in clotting, and the discovery of human blood groups in 1901. This precarious in-between stage is reflected in the scene describing the character Lucy's blood transfusion, a procedure necessitated by Dracula's secret nightly feedings. When choosing a compatible blood donor, the two doctors treating Lucy never mention blood type. A, B, and O are a few years off. This being Gothic fiction, the desired sex of the donor is also never questioned. "It is a man we want," Dr. Van Helsing states, implying in these six syllables all sorts of manly virtues, such as vigor. A spirited game of scissors-paper-rock, I imagine, is waylaid by the arrival of the ideal donor, Arthur. The youngest, strongest, and "calmest" of the three men, Arthur is also madly in love with Lucy, a quality that seals the deal.

Next, Stoker makes a little stumble. He has Van Helsing happily point out that Arthur is "of blood so pure that we need not defibrinate it." Only in a work of fiction would this be considered an advantage. Blood that does not produce fibrin is blood that doesn't clot. A real-life Arthur would've been suffering from a disorder analogous to hemophilia and clearly would not have been a doctor's first choice when selecting whose vein to slice open. In the story, however, the physicians are glad to avoid the sticky problem of coagulation, which I can certainly appreciate. Upon exposure to air, blood at the site of an injury immediately begins to clot, or congeal. The body is trying to self-seal. Platelets (so named for looking like tiny plates) converge on the site, clumping together and simultaneously secreting chemicals that turn the blood-borne protein fibrinogen into long, sticky threads. Red and white cells get caught in the webbing, and the clump becomes a clot. At a wound site, clots are lifesaving. But within the circulatory system, a clot may lodge in a blood vessel (this is called an embolism) and cause stroke or death. While anticoagulants prevent clots from forming during transfusions today, such agents did not exist during the era portrayed in *Dracula*—hence the need to defibrinate.

Defibrination was a crude but clever process developed in

the 1820s and used up through the introduction of anticoagulants in the 1920s. Wildly varying methods arose, but each took time, so I can see why Bram Stoker, if only to keep the scene moving, had Dr. Van Helsing skip this step. One method involved collecting the donor's blood in a bowl, whipping it with a wire eggbeater, then filtering out the clots through a stretch of gauze. Even simpler was allowing the collected blood to settle for several minutes and then scooping out what congealed. Sometimes, too, the blood was twirled in a flask containing glass beads around which clots would form. These methods were not foolproof—bacteria entered the process, and clots slipped through—but transfusions had become safer, if only just. (To be fair, they did represent a vast improvement over the previous treatment for blood loss, bloodletting. Up until the 1820s, for example, a woman with uterine hemorrhaging following childbirth was commonly bled. Now, there's a horror story.)

Once Lucy is drugged up, Van Helsing proceeds. Arthur lies next to his fiancée while the doctor removes from his bag the necessary instruments—what he calls the "ghastly paraphernalia of our beneficial trade." Although Stoker doesn't linger long on the procedure, details suggest that his fictional doctor may be performing an actual type of transfusion that, at the time *Dracula* was written, would have been highly experimental, a direct artery-to-vein transfer. This method, considered promising in animals, made a short-lived leap into human use in the late 1890s. You'll soon see why it so quickly came and went. Typically, the blood donor's radial artery (one of the two main arteries in the forearm) was exposed, distended, tied off or clamped, sliced open, and then either sewn directly to the recipient's similarly exposed vein or connected to it via a small metal pipe. Bodies had to be aligned just so. Once the clamps were removed, the donor's heart literally served as a blood pump. But, like gassing up the SUV without benefit of a meter, the transfer was difficult to measure. Too much blood? Too little? In some cases, the blood donor was simply weighed before and after "the operation," to borrow Stoker's apt phrase, and the difference used to estimate the volume taken. (A bit late, no?) In *Dracula*, Van Helsing demonstrates an alternate method. For several tense minutes, his gaze darts among Lucy, Arthur, and his pocket watch, used to time the flow. Once some unexplained threshold is reached, he announces, "It is enough."

It *is* enough. A bloom returns to Lucy's cheek, and the possibility of her heart being overwhelmed by too much blood has passed. Arthur, who could've suffered excessive blood loss, is shaky but also fine. While the medical dangers have been avoided, however, a supernatural one remains. Dracula, unbe-

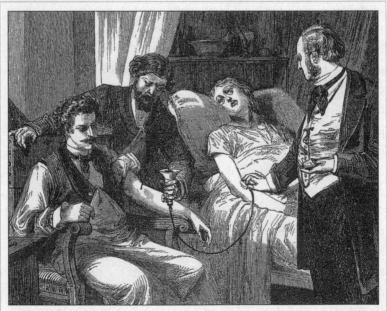

Performing "the operation" of transfusing blood

knownst to all, continues to feed on Lucy. Over the next ten days she receives three more transfusions, each helping her less. The heroic efforts fail to save Lucy's life. Dracula drinks her to death, and Arthur's dream of marrying the beautiful young woman is dashed. Heartbroken, he consoles himself with the fact that a consummation of sorts had taken place: "The transfusion of his blood to her veins had made her truly his bride."

But alas, Dracula's bride as well.

Dr. Van Helsing, demonstrating a knack for good guesses, concludes that poor dead Lucy is now one of the Un-Dead. A trip to the cemetery later confirms his hypothesis. Lucy lurks among the tombstones, feeding off a child. Transformed by Dracula's blood, she is "like a nightmare of Lucy," her sweetness "turned to adamantine," her purity to "voluptuous wantonness." She approaches Arthur, who has joined the doctor: "My arms are hungry for you," she purrs. A brandished crucifix forces her retreat.

In the scene set the night thereafter, Stoker was clearly intent on steaming his audience's reading glasses. The lid of Lucy's coffin is lifted and her sultry form is revealed. With meaty stake in hand, Arthur awakens her with a great thrust. Lucy writhes, moaning behind deep red lips. Her "body shook and quivered and twisted in wild contortions." Arthur plunges again, drawing blood. "He looked like a figure of Thor," hammering, "driving deeper and deeper." But for the reality of the stake that finally pierces her heart, she seems to be enjoying this. Lucy gives a final shudder, then is still. If Dracula is even aware of her demise, he doesn't give a damn. Once he's "turned" a woman, he loses all interest in her and moves on.

. . . .

JUST BENEATH THE SURFACE, BARELY, *DRACULA* IS A CAUTIONARY TALE about the evils of submitting to one's darkest desires. This reflects Bram Stoker's Victorian and Christian sense of morality. At the same time, the writer was savvy enough to know that excessive finger-wagging does not a bestseller make. By making the sex metaphorical, he was able to push against the edge of propriety, just this side of objectionable, without sullying either his own or his upstanding characters' reputations. Lucy, for instance, dies a virgin, despite her having been, forgive the inelegance, penetrated countless times in various ways—fanged by Dracula, poked by doctors, infused with donor blood, staked by her fiancé. In the end, as death releases the vampirism from her body, Lucy returns to a picture of purity, her original self. What Stoker accomplished with words reminds me some of Alfred Hitchcock's approach to horror while shooting his 1960 film *Psycho*. When asked after its release why he hadn't used color film, which was, of course, available at the time, Hitchcock replied, "Because of the blood. That was the only reason." Had he shot the infamous shower stabbing scene in Technicolor, the studio censors would've done their own slashing. "I knew very well I'd have the whole sequence cut out," he said. In black and white, though, he could get away with, well, murder.

With *Dracula,* Bram Stoker was determined to create a substantial work of literature that would make his name. In the dozen years before he started his first draft, he'd dashed off ten pieces of fiction, including another novel. As to their reception, a phrase comes to mind: It's a good thing he kept his day job. As the secretary and business manager for Henry Irving, the foremost Shakespearean actor of the time and a world-class prima donna, Stoker had to squeeze writing into moments snatched between beck and call. He slowly built the character of Dracula, who, though he would become literature's most enduring vampire, was not in fact the first. Three had come before, and Stoker culled important elements from their tales. Dracula's seductive ways, for example, owed a debt to the lusty female vampire of *Carmilla* (1872), a Gothic novella written by fellow Irishman J. Sheridan Le Fanu. (Le Fanu was Stoker's boss at a Dublin newspaper at the time of *Carmilla*'s release.) Dracula's black cape, the wooden stake, and the notion that vampirism could be passed to others through blood exchange were details borrowed from James Malcolm Rymer's *Varney the Vampire, or The Feast of Blood* (1847), a 750,000-word saga that had originally appeared as a "penny dreadful" serial. Finally, the delicious casting of Dracula as a nobleman, a count, living in the midst of and feeding on members of high society descends from fiction's very first vampire, Lord Ruthven, who appeared in John Polidori's short story "The Vampyre" (1819).

The story behind Polidori's story is far better than his final product. Twenty-year-old Dr. John Polidori, a British physician with literary aspirations, was sharing a lakeside villa near Geneva, Switzerland, with the poet Lord Byron, who'd fled London due to debt and allegations of an extramarital affair between him and his half sister. This was the summer of 1816. The two men had what one might call a give-and-take relationship: Byron freely took the opiates Polidori could legally obtain and, in return, gave the doctor the opportunity to orbit in literary circles. During several weeks in June, the gentlemen were joined by three invited guests: England's leading poet, Percy Shelley, his young lover Mary Godwin, and her stepsister, Claire. Claire, as had and would several other women, was carrying Byron's illegitimate child. The two immediately had a tiff, and Byron from then on would speak to Claire only in the presence of the others. Perhaps the pets in residence picked up on the tension. Percy recalled that "eight enormous dogs, three monkeys, five cats, an eagle, a crow, and a falcon" all moved freely about the house, "which every now and then resound[ed] with their unarbitrated quarrels." Bad weather further tested everyone's nerves. A series of fierce rainstorms kept the group housebound for days. One night Byron and company, desperate to fill time, resorted to reading aloud from a collection of French translations of old German horror stories, perhaps left behind by a previous renter. (As a contemporary analogy, I imagine a klatsch of socialites reading to each other from the lyric sheet of a rap CD.) The stories were horrifically bad. Byron felt that he and the others could surely do much better, and, as an amusement, issued a challenge: "We will each write a ghost story." Now, the two people one would think most likely to produce something magnificent didn't get very far: Byron and Shelley both had ideas, but quickly abandoned their efforts. But not eighteen-year-old Mary Godwin. An idea came to her in a dream, and she began working feverishly on what would two years later be published under her married name, Mary Shelley's *Frankenstein* (1818). As for the doctor in the group: "Poor Polidori," Mary would later recall, "had some terrible idea about a skull-headed lady." It seems that Polidori fizzled not just creatively but socially, too. By summer's end he and Byron had severed their relationship, igniting an enmity they would both carry through the rest of their lives. Polidori, still hoping to be a writer, took the idea that Byron had discarded—the skeleton of a vampire tale—and began adding meat to the bones. Out of spite, Polidori fashioned the villain of the reworked piece after Byron. Enter the bloodsucking aristocratic fiend "Lord Ruthven." Even in this name, though, Polidori wasn't original. He'd borrowed it from a roman à clef written by one of Byron's ex-lovers. Thus, in an exhalation of venom, "The Vampyre" was born.

Seventy years later, once Bram Stoker had given himself the

challenge of writing a vampire classic, he, too, borrowed a name for his villain, although he took from history, not fiction. Vlad Dracula (1431–1476) was a prince born in the Transylvania region of Romania. *Dracula* came from his father's nickname *Dracul,* meaning "the dragon"; the added *a* indicated junior status. Vlad, "son of the dragon," sometimes translated as "son of the demon," would emerge as a leader on the Christian side of the long-standing war against the Muslim Turks. He excelled in cruelty. He concluded one battle, for instance, with the command that the thousands of captive Turkish soldiers be impaled, a slow and horrific way to die, a public butchery also meant to torment the survivors. By this point Dracula had earned a new sobriquet: Vlad Tepes, or Vlad the Impaler. Tales of his viciousness took on such life that, after he was slain by a Turkish assassin, the sultan of Constantinople ordered that Vlad's head be staked and displayed. *Come, believe your eyes, the demon is dead.*

In co-opting the name *Dracula,* Stoker was aiming not to model his character after the historical figure but to evoke a kindred spirit of evil. Stoker may have employed a similar tack regarding another member of Transylvanian nobility, Elizabeth Bathory (1560–1614), although here a suggestion of real-life vampirism is not out of the question. Bathory, it's reputed, regularly bathed in human blood, which she believed would preserve her youth and beauty, as Raymond McNally details in *Dracula Was a Woman* (1983), a biography of the so-called blood countess. Coincidentally—or, then again, maybe not— Count Dracula, over the course of twenty-seven chapters, grows ever more youthful as he drinks his victims' blood, a theme that hadn't occurred in earlier vampire tales. Because this was so Bathory-like, McNally contends that Stoker had indeed been inspired by her and points out that the first account in English of the Bathory case was included in a book Stoker used as a reference, a nineteenth-century encyclopedia of the supernatural. But did Stoker, I wonder, even read this entry? Couldn't he have just plucked the de-aging idea from his imagination? Stoker scholars and vampire enthusiasts hotly debate this question. Speculation notwithstanding, Bathory's legend is rich in its own right. Her victims were peasant girls, preferably virgins, either hired as servants or kidnapped outright. However obtained, they all ended up in the castle cellar, the location of Bathory's torture chamber, where they were eventually exsanguinated. Of several gruesome methods, one involved being locked inside a spike-lined spherical cage that was hoisted to the ceiling, then rocked so that the girl was pierced again and again. The countess stood naked beneath, in the shower of warm drippings.

Not to excuse her behavior in any way, but some historical context might be helpful at this point. The use of blood in one's beauty regimen was not unheard of in the sixteenth century. To prevent wrinkles, wealthy women of the Renaissance would rub their faces each morning with the Kiehl's moisturizer of its day, the blood of doves. As for the use of virgin blood, there, too, were numerous precedents. Aztec priests in the fifteenth century, to cite one example, sacrificed virgin girls as offerings to their primary deity, the corn goddess. In Europe

Portrait of Elizabeth Bathory, the so-called blood countess, at age twenty-five

during the Middle Ages, it was believed that physical illness, thought to be brought on by sin, could be washed away using "innocent" virgin blood, though the donor did not need to be killed. Variants on this thinking continued as late as the fifteenth century, according to medical historians, at which time a draft of the blood of a young person, for instance, might be prescribed for the rejuvenation of the aged. I suppose it goes without saying that none of these factoids was ever brought up in Elizabeth Bathory's defense.

Arrested on December 30, 1610, the fifty-year-old countess was charged with committing what a panel of judges called "an almost unbelievable number of murders." Through surviving court documents from her two trials, it's possible to sift a handful of facts from the voluminous legend that has since engulfed Elizabeth Bathory. The countess was not present at either trial (she'd been placed under house arrest in her castle), but her four closest servants, charged as accomplices, were brought before the judges. Previously tortured, the servants, one by one, ratted out their boss. The body count was thirty-six, thirty-seven, or fifty-one girls, depending on whom you believed. Another witness, not charged, claimed the number was much higher. She testified to what she'd heard secondhand: A castle servant had found among Bathory's possessions a handwritten list of victims, 650 in all. This smoking gun, however, was never introduced into evidence. Neither was a word said regarding Bathory's bloodbathing. Nonetheless, snippets of the transcript remain chilling: "The countess stuck needles into the girls." "She bit out individual pieces of flesh . . . with her teeth." She "attacked the girls with knives" and "beat them so hard that one could scoop up the blood from their beds by the handfuls." If the servants had been hoping for leniency, they were sorely disappointed. Three were sentenced to execution—one was beheaded; two were burned alive after being de-fingered—and the fourth to life in prison. Bathory, too, was given a life sentence, though, as a concession to her noble lineage, this meant confinement to a small room in her castle, the windows and doors of which were bricked in, save for a slot for food. Till her death three years later, she maintained her innocence.

In terms of sheer villainy, one can easily imagine how the stories of Vlad and Elizabeth may've inspired Bram Stoker's Dracula. But to tap into that essential cringe factor, the novelist turned to the animal kingdom. To the *Desmodus rotundus* in particular—the vampire bat. That Stoker pored over a description in the 1823 edition of *Anecdotes of the Habits and Instincts of Animals* prompted my own poking into present-day sources. Apart from its repellent appearance—the beady eyes, horsey ears, and piggish snout—what makes the vampire such a nauseating bat is its signature mode of feeding: A nighttime hunter, it lands on the ground a few feet from its victim, usually a sleeping cow or horse, and skitters forward on all fours. It's said that sleeping human beings are sometimes its prey, so you may want to splurge on the extra-strength mosquito netting next time you're in Central or South America, the species' native habitat. Powerful hind legs aid the sparrow-sized mammal in scaling a dangled arm, leg, or tail. The bat then sinks its razor-sharp canine teeth into a fleshy area such as the neck, having first licked soft the spot. Its saliva, which contains an anti-clotting enzyme, keeps the blood flowing while the vampire sucks. (So potent is this superdrool that scientists have synthesized the anticoagulant into a powerful blood-thinning medication called Draculin, appropriately enough.) A good thirty minutes of nightly feeding meets a bat's necessary daily intake; the vampire survives wholly on blood. The bat's bite can also spread disease (rabies, for instance), and although Stoker doesn't expressly say so, this is also how Dracula transmitted his contagion. Vampirism is an infectious disease in which evil is the pathogen. With each new bite, one's essence is overwhelmed, one's blood overpowered.

BRAM STOKER TOOK GREAT CARE TO CLOAK HIS VAMPIRE TALE IN THE guise of realism. The more authentic and contemporary his fictional world, he rightly believed, the more genuine the reader's fright. Hence his characters used such newly invented devices as a recording phonograph (an early-model tape recorder), a portable typewriter, and a Kodak camera. Likewise, a trip to Transylvania taken by one of the book's heroes followed actual train timetables. Elsewhere, landmarks and locations were drawn from real life, as were certain events. A ship that had beached near where a vacationing Stoker wrote portions of *Dracula,* for example, found its way into the plot. When he was uncertain of details, he turned to experts—to his older brother, Thornley, for instance. A prominent surgeon in Ireland, Thornley Stoker vetted the final manuscript before Bram sent it to the typesetter, double-checking the blood transfusion scenes, in particular, to make sure of their accuracy.

Whereas Stoker wrapped his myth in truths, modern-day scientists have worked to expose the truths behind the myth, posing, for instance, the fascinating question, Might there have been a medical basis for the allegorical disease of vampirism? The answer is a resounding yes. It's a blood disorder called porphyria.

In its mildest symptomatic form, porphyria is not at all vampiric—at worst, an extra sensitivity to sunlight may cause your skin to blister. But in rare cases an untreated victim may indeed look like one of the undead: Your coloring takes on a deathly pallor due to severe anemia; your lips erode and gums recede, making your teeth—the eyeteeth, in particular—appear longer, more fang-like; and sunlight turns your affected flesh caustic, causing your facial features to dissolve and fingers to be eaten away. The lesson you learn quickly is, daylight is deadly. That undiagnosed cases of porphyria may've first planted suspicions of vampirism hundreds of years ago, perhaps as far back as the twelfth century, was originally suggested by a Canadian biochemist in 1985. Little could Dr. David Dolphin have imagined as he stepped up to the conference podium that day in May the media monster he'd unleash. What assured he'd grab headlines was his matter-of-fact contention that victims may have been driven to drink blood to relieve their symptoms. After an initial fireball of attention, Dr. Dolphin's hypothesis has since taken on a life of its own, especially on the Web, not unlike the legend of Elizabeth Bathory.

Porphyria is triggered by a flaw in the cellular machinery for producing heme, a crucial element of the blood's oxygen transporter, hemoglobin. One of the steps in assembling heme involves the introduction of dark red pigments called porphyrins (from the Greek for "purple"). When the system is flawed, you end up with too much porphyrin and not enough heme. The porphyrin pigments backlog, building up in the skin, teeth, bones, and organs, causing a host of symptoms depending on where the accumulations occur. Your teeth may turn a dirty brown, for instance, and pain may settle into your limbs and back. (That sufferers can be extremely sensitive to sunlight made more sense once I learned that porphyrins are an ancestral sibling to chlorophyll, though, of course, the light-activated process of photosynthesis in plants isn't destructive.) While toxins such as drugs, alcohol, or chemical poisoning can bring on porphyria, the illness is mainly hereditary in origin.

It's now known that the infamous British king who

reigned during America's war for independence, George III (1738–1820), suffered from acute intermittent porphyria (AIP), one of eight distinct forms of the disease. As is typical of AIP, the king's illness manifested most notably in neurological symptoms: seizures, hallucinations, and bouts of mania and paranoia that would last for days or weeks at a time, then vanish, with long remissions in between. That his malady was porphyria, not "madness," as was believed during and long after his reign, would be unknown today were it not for a peculiar fact of royal life: As the monarch, George was subject to daily visits by physicians, who chronicled his every symptom. From these surviving documents, modern-day British researchers have gleaned conclusive evidence for a posthumous diagnosis of AIP. Aside from jottings regarding the characteristic mental paroxysms, which began when George was in his twenties, the clincher was this notation: "His Majesty has passed . . . bloody water," by which was meant discolored urine, elsewhere described as "bluish," dark and "bilious," and having left "a pale blue ring" around the specimen flask. All are telltale signs of excessive porphyrin production, as are severe abdominal pain and muscle weakness. The medical records also show how royal protocol must have frustrated the doctors, who could never speak unless first spoken to. When King George was at the peak of his delirium, entire visits passed in silence, as on one day in January 1812: "His Majesty appears to be very quiet this morning, but not having been addressed we know nothing more of His Majesty's condition of mind or body than what is obvious in his external appearances." Perhaps this explains the preoccupation with the kingly pee.

Having exonerated George III of madness, the British researchers then posed the next logical question: Since AIP is always hereditary, who else in his bloodline carried the disease? By combing through historical accounts and medical records—a search abetted, once again, by the fastidious description of urine samples—they were able to trace the disorder through thirteen generations, spanning more than four hundred years. Among his ancestors, fifteen were identified as sufferers and/or carriers, beginning with his father and going back to Mary, Queen of Scots (1542–1587), where the porphyria paper trail ended. As with George, retrospectively diagnosing Mary with the illness allows for a radically revised view of her reign. Mary, whom the researchers called "one of the great invalids of history," was so sick so often that her opponents accused her of using hypochondria, to borrow the modern term, as a political ploy. Conversely, her enemies were accused on one occasion of poisoning her, an episode now ascribed not to foul play but to the fickleness of genetics.

Such a rare disease as porphyria would never have manifested in so many family members were it not for the strict controlling of the bloodline through intermarriages, a phenomenon not unique to royal dynasties. A similar kind of inbreeding occurred within isolated and remote communities during the Dark Ages, for example, and, in these shallow gene pools, recessive traits could flourish. Hence, as biochemist and medical writer Nick Lane postulates, a type of porphyria that is among the rarest today—congenital erythropoietic porphyria (CEP), the disfiguring, vampiric form of the disease described earlier—may've once been relatively common in spots of Eastern Europe now recognized as the cradle of vampire myths, the valleys of Transylvania. Assuming this to have been the case, it's easy to imagine how the corpse-like appearance and odd behavior of sufferers may have given rise to whispers of vampirism; how, within these enclaves, certain folk remedies would've been embraced; and how, over time, the rumors and remedies would have gradually evolved into legend. Garlic is a good example. It's now well known that certain chemicals in garlic can exacerbate porphyria symptoms, a lesson that Transylvanian sufferers may have had to learn through painful experience. Little agility is needed to make the next leap, to imagine how a sufferer's way of averting a flareup could mutate into a superstition among the healthy for preventing the disease and then into a means of warding off a vampire attack. Likewise, the genuine need to avoid the sun could have transmuted into the dramatic literary convention of the bright light of day turning a vampire to toast.

In view of all this, the hypothesis put forth by David Dolphin doesn't sound so far-fetched: that, hundreds of years ago, victims of this most heinous form of porphyria may have self-medicated by drinking human blood. In a sense, this is evocative of earlier thinking, such as the belief in ancient Rome that a swallow of gladiator blood could cure epilepsy. Provocative, yes, but scientific, no. In Dr. Dolphin's theory, however, science concedes. The severe anemia caused by CEP leaves sufferers with dangerously low levels of heme in their circulatory systems. In medical terms, a heme deficiency is an iron deficiency, which is why the modern treatment for this rare form of porphyria is regular transfusions of blood. Though not recommended, a patient could instead be given a straw. The heme molecule is robust enough to survive digestion and will make its way into the bloodstream.

When I consulted a nutritionist on this final notion, I was so preoccupied by the repulsive thought of ingesting blood that I was startled when she stated the obvious: "We eat blood all the time. It's in our meat, in all the animals we kill for food." Mary Kay Grossman is a registered dietician and coauthor of the bestseller *The Insulin-Resistance Diet.* "In our culture," she continued, with the exception of kosher diets, "we don't drain the blood off. If you cook it, it's not unhealthy to eat blood, and it doesn't lose its nutritional value." In fact, Grossman explained, some cultures, such as the Masai of Kenya and Tanzania, subsist entirely on blood and milk—cow's blood, that is. "They milk the cows, then puncture the throat, and drain the blood off." (The cows survive, by the way.) The Masai then mix the two and drink it fresh, she added, or give it a few days to ferment. "They live in an extremely dry climate where it's almost impossible to grow anything, so the blood supplies iron and the milk is a major source of protein."

Raw animal blood is a central part of the diet of other pastoral groups in eastern Africa, I later learned, including the Karimojong of Uganda, but, globally, cooked animal blood as a main ingredient in traditional dishes is far more common. The Inuits with their seal's blood soup, for example. The Tibetans with yak's blood cubes, a snack of reduced yak's blood served with sugar and hot butter. And the English with their black pudding, a baked then fried concoction of pig's blood, bread cubes, skim milk, beef suet, barley, oatmeal, and mint. You can even taste your way through every region of France through local interpretations of *boudin noir,* "blood sausage." *Larousse Gastronomique,* the classic French encyclopedia of Continental cuisine, describes sixteen variations, following the basic recipe of equal parts onions, pork fat, and pork blood.

King George III was likely never fed such savory dishes when he was ill. Rather, according to historians, he was often straitjacketed, tied to a bed or chair, and subjected to a varying regimen of vomits, purges, blisterings (the placing of hot coals on the skin in order to draw "bad humors" to the surface), cuppings, bloodlettings, and leechings. Heaven help the king. Perhaps some degree of solace can be found, however, in the fact that the last three forms of treatment would have, in theory at least, helped the ruler. While anemia can be relieved through adding blood, removing blood will quickly reduce the level of porphyrins in circulation. In fact, some types of porphyria are currently treated through phlebotomy, the modern-day counterpart to bloodletting, as is the more common hereditary blood disease, hemochromatosis, in which a dangerous excess of iron in the blood must be decreased through regular blood draws. (A curious side note: Once this heme-heavy blood is collected, it's routinely destroyed, even though—assuming the donor with hemochromatosis is otherwise disease-free—its iron-richness is exactly what many ER patients need.)

Four of King George's sons are believed to have had porphyria, including the heir apparent, George IV, whose wife (a first cousin) and daughter, Charlotte, were also afflicted. It's quite possible the illness later led to Princess Charlotte's death during childbirth at age twenty-one (her son was stillborn), a tragedy that precipitated a regency crisis: The king, who by this point was blind, enfeebled, and nearly eighty, now had no legitimate heir beyond George IV. Moving swiftly, he arranged marriages for his three eldest sons, and each produced a child in the year 1819, one of whom, Victoria, would be crowned queen of England at age eighteen. Although Victoria, who also married a first cousin, Albert, was spared porphyria, she introduced another devastating blood disease into the British royal family, one that would eventually taint the ruling houses of Spain, Germany, and Russia: hemophilia.

Shemophilia

BLOOD NATURALLY SEPARATES. REMOVED FROM THE constant "stirring" of the circulatory system and collected in, say, a test tube, blood settles into our tricolor hematological flag, amber, white, and red. The band at the top is plasma, the liquid in which the cells of the blood are normally suspended. Next is the narrowest stripe and the palest, a blend of white cells and platelets. And beneath these, the bog of burgundy-colored red cells, the heaviest of which, the deep red, almost black corpuscles laden with waste, have sunk like sediment at the bottom of a pond.

Curiously, what some would call the defining quality of blood—its redness—does not in fact contain the defining quality of humanness, DNA; red cells are "dumb" cells, devoid of a cellular "brain," a nucleus.

The color blue is not part of the mix, although the long-lived phrase *blue blood* deserves a little deconstruction. Its etymology begins with *sangre azul,* a term that arose out of a medieval case of xenophobia. During the Moorish occupation of Spain, members of the oldest and staunchly Christian families of the Castile region claimed they were superior for never having intermarried with the darker-skinned Muslim invaders of their country. The proof of their blood's purity was as near as a forearm. A Castilian need only point to the tributaries of blue visible through his or her white skin. What they called *sangre azul* we'd now call an optical effect, deep purple blood seen through pale purplish veins seen through an epidermal scrim.

By the time *blue blood* crossed into the English language in the 1830s, concurrent with the beginning of Queen Victoria's reign, it had shed the racial connotation and become synonymous with society's upper crust. Within this rarefied meringue of British blue-bloodedness were further distinctions: the gentry class, the aristocracy, and, at the top, the royal family. Royalness, too, had its own degrees. To be of "morganatic blood," for example, meant that one of your parents was of pure royal extraction—that is, "of the blood," the bluest of the blue—and the other was not. Perhaps your mother and father had married not as a dynastic stratagem but out of love—what folly! Entering into a morganatic marriage came with a price: the forfeiture of your children's right to succession. Compared to other sovereigns of her era, Queen Victoria was far more accepting of such unions. A sterling example of this magnanimity came in the spring of 1866. When informed that an obscure Austrian prince wished to marry one of her cousins, Victoria dismissed the many objections regarding the gentleman's unequal birth and gave her full blessing. What's more, upon first laying eyes on this tall, strapping man, the queen saw not the penniless military officer but a solution to a problem, one the prince's physical presence brought to the fore. What the queen would never admit publicly was her deep concern about the quality of her family's blood, made, as she described it, "so lymphatic," generation after generation, by all "that constant fair hair and blue eyes." Prince Teck was all things but with jet-black hair and dark good looks. Shortly after meeting him, the queen wrote to her eldest daughter, Vicky, a lifelong confidante, who was grown with fair-haired children of her own and ensconced as the crown princess of Prussia. Oh, Victoria lamented, "I do <u>wish</u> one could find some more black eyed Princes or Princesses for <u>our</u> Children!"

It's hard to imagine Vicky's reaction as she read the rest of her mother's words, which, in just a handful of sentences, move from despair to envy to a rising franticness (note how even the underscorings escalate): "I can't help thinking what dear Papa [the queen's deceased husband] said—that it was in fact a blessing when there was some little <u>imperfection</u> in the <u>pure</u> <u>Royal</u> descent and that some fresh blood was infused." Here the queen paused for a moment, asking her daughter to excuse this "somewhat odd letter," before diving back in: "It is <u>not</u> as <u>trivial</u> as you may think, for darling Papa—<u>often</u> with vehemence said: 'We <u>must</u> <u>have</u> <u>some</u> <u>strong</u> <u>dark</u> <u>blood</u>.' "

This letter reads as if a mother's intuition had sharpened in grandmotherhood, but at the time Victoria would've had no idea of the troubles to come. The hemophilia that would eventually touch sixteen family members had, in 1866, manifested solely in Victoria's youngest son, thirteen-year-old Leopold. A strange aspect of the disease is that while females carry the defective gene that prevents proper clotting, in general only males develop the disorder. In other words, it stays hidden in a woman until it shows up in a son. By scanning Queen Victoria's family tree, medical historians have established that two of Leopold's five sisters, Alice and Beatrice, were carriers. There's also no doubt that his mother introduced hemophilia into the royal bloodline.

How Victoria got it is something of a puzzler. Tracing back her ancestry, no red branchings of the disease appear, which leads to three possibilities. The traditionally held view is that it was caused by a spontaneous mutation (this occurs in about 30 percent of hemophilia cases). Second, that, against the odds, Victoria's mother, maternal grandmother, maternal great-grandmother, and so on were carriers whose sons never suffered the disorder. Or, third, the most sensational possibility, that Victoria was illegitimate. Genetics has fueled this particular speculation. Given that every daughter of a male hemophiliac (and a normal female) is a carrier, perhaps then, as a pair of British scientists postulated in the mid-1990s, Edward, Duke of Kent, was not her biological father. (As a *Newsweek* headline blared, "Was Queen Victoria a bastard?") Well, maybe, maybe, and maybe, although I'm leaning toward (1), spontaneous mutation.

To her dying day the queen refused to believe that the disease came from her side of the family. It is also considered unlikely that she was ever fully briefed on the cause and patterns of the disorder, even though a fairly sound clinical description had been established at the dawn of the nineteenth century. In Leopold's early childhood, no obvious flags went up. He was her most "delicate" son, Victoria admitted, born tiny, and less graceful than his three older brothers. She blamed his frequent bruising on clumsiness and, in her ignorance of Leo's true condition, was ofttimes impatient and critical. Sickly, pallid, and frail, the boy was an embarrassment. At the time of his fifth birthday, though, a family walk, a skinned knee, and a cut that would not stop bleeding forced Victoria to face the reality that her son was a "bleeder." With that, the queen made an emotional turn-around, and a well-intended overprotectiveness set in. She drafted an all-staff bulletin of sorts regarding her son. Forthwith and henceforth, all active play with other boys would be denied the lad, and "all the <u>essentially</u> English <u>notions</u> of '<u>manliness</u>' must be put out of the question." His tutor must never leave Leo

Leopold, age nine, with his mother, Queen Victoria, April 1862

unattended, and a long list of activities became restricted. But of course, tell a child "don't" and he'll naturally be tempted to "do." When Leopold was eight, to cite one example, he somehow managed to ram a steel pen through the roof of his mouth.

Stitches didn't work well for a hemophiliac because, obviously, they just introduced more holes. The alternative was cauterizing, a method that essentially melted closed a wound, using either a caustic substance or a red-hot brand. I can only hope that Leopold was well anesthetized when this treatment was inflicted. When Victoria described this incident in a letter to her daughter Alice, her words seemed to stumble, as if numbed: "The fear was—the bleeding could not be stopped and then—you know he could not have lived."

That Leopold survived into his early thirties is medically remarkable. Most bleeders in his day never made it into adolescence. In fact, individuals with severe hemophilia faced a similar mortality rate well into the 1960s. To answer definitively why Prince Leopold lived as long as he did, however, would require something that simply no longer exists: a sample of his blood. With it, a modern hematologist would be able to find exactly what was missing. Without it, a good guess is still possible. But first, some basics are in order.

The simplest way to describe coagulation is to say it's a complex process in which blood turns from a liquid to a solid. As many as twenty different blood proteins take part in this coordinated effort—what one scientist with a touch of the poet named "the clotting cascade." Thirteen of these blood proteins are called factors, and a deficiency of any one results in a different clotting or bleeding disorder. To picture the prevalence in the general population, imagine a twenty-thousand-seat sports stadium filled to capacity and split evenly between the sexes. Four hundred attendees, men and women alike, would have the most common bleeding disorder, von Willebrand's disease, in which a deficiency of the von Willebrand factor keeps platelets from clumping properly at the site of an injury. Just two attendees, men only this time, would have hemophilia A, or classic hemophilia, which is caused by a lack of factor VIII. To find a single gentleman with hemophilia B, two additional stadiums are needed (one in thirty thousand men have this deficiency in factor IX). Now on to a bigger challenge. To find a woman with classic hemophilia, one should probably scrap the whole stadium idea and consider instead the entire population of the United States. One in one hundred million citizens is this rarest of the rare, a female hemophiliac. Its infrequency in women is simply a matter of genetics, the female XX versus the male XY. The recessive gene for hemophilia rides the X chromosome. In a female, if one X is defective, the other can normally compensate. It would take the pairing of two defective X's for the condition to develop.

In males, the first sign of a serious bleeding disorder often comes when a baby boy is circumcised, a danger that has been recognized since ancient times. In the Babylonian Talmud, the collection of Jewish rabbinical laws written between the third and sixth centuries, it was declared that a newborn son would be excused from circumcision if two brothers had previously died from the procedure. From a modern standpoint, this law calls to mind a common misconception that should be stanched forthwith: that a hemophiliac with, say, a minor wound will never stop bleeding. The lack of factor VIII or IX doesn't mean your blood won't ever coagulate; the woodwinds may be absent, so to speak, but the orchestra still plays. Other components of the clotting cascade continue to do their job. The problem is, you clot more slowly and hence bleed longer. How long depends largely upon how much clotting factor is present in your blood. Analyses of blood can calculate the speed of your clotting as well as pinpoint which factor is deficient and in what quantity. A person with mild hemophilia, for instance, will have, at best, only half the clotting activity of a healthy person. By contrast, in severe hemophilia, the amount is less than one one-hundredth of normal. Once these calculations are determined, treatment is fairly straightforward, at least in theory. You simply inject or infuse the missing substance. And of course the earlier in a person's life that these levels are assessed, the better.

Regarding Leopold, it is possible to piece together details of his condition through his words, for the young man's personal correspondence lays bare his suffering. A well-understood medical consequence of severe hemophilia is spontaneous internal bleeding into the joints and muscles, which balloon with blood, becoming excruciating and crippling. Leopold clearly had this. In a letter to his sister Louise begun June 6, 1870, he barely got past the "Dearest Loo" before having to stop, so fierce was the pain. He couldn't continue until four days later: ". . . At this moment I am in agonies of pain; my knees get worse daily and I get more desperate." Despite the limited relief offered by the treatments of the day—bed rest, ice packs, and, only as a last resort, morphine—Leopold, seventeen at the time, leavened his note with a bit of gallows humor. "If this continues long I shall soon be driven to Bedlam [by which he meant the loony bin], where I shall be fortunately able to terminate a wretched existence by knocking out my brains (if I have any) on the walls; that is the brightest vision that I can picture to myself as a future. . . ."

Signed, "Your wretched brother, Leopold."

If humor helped him rise above the pain, an intense pursuit of academics also provided an escape. By this point in his life, the Scholar Prince, as he would be known, had become well versed in Shakespeare and fluent in several languages. Over his mother's objections, Leo attended Oxford University, the first in the family to do so. His gumption is all the more remarkable when you factor in that he also had epilepsy. After graduation he became one of the queen's most trusted political advisers, gaining the title Duke of Albany. At the same time, he wanted to establish a life for himself apart from his mother and longed to marry. In 1882, at age twenty-nine, Leopold made both himself and the queen happy by wedding Princess Helena of Waldeck, sister of the Dutch queen. The happiness would last just two years. Shortly before the birth of his second child, Leopold took a spill—what for a healthy person would've merely meant a bump on the head. He died a few hours later from a brain hemorrhage. Upon receiving the news of her son's loss to hemophilia, Queen Victoria, now sixty-five, set down in her journal three devastated words, "Am utterly crushed."

At this stage of her reign, Victoria was becoming known to the world as "the Grandmama of Europe," so called because so many of her children and grandchildren had married heirs to thrones throughout the Continent. The Grandmama of Europe was also the hands-on grandmama of six grandchildren, in whose raising she'd taken an active role since her daughter Alice's passing from diphtheria in 1878. When these children approached marrying age, she did more than matchmake. As with many of the other royal marriages she'd helped broker, these unions would expand the family power base. They would also spread hemophilia geographically. In 1888 granddaughter Irene was married to a first cousin, Prince Henry, thus bringing hemophilia into Prussia. In 1894 granddaughter Alexandra was married to Czar Nicholas II, tainting the Russian imperial family, the Romanovs. The 1885 marriage of Victoria's youngest child, Beatrice, also bears noting here, for it brought the stain into the German royal bloodline. A daughter of this union would then go on to introduce hemophilia into the blue blood of Spain. Closest to home, Leopold's daughter would perpetuate the family legacy in her marriage to a British nobleman.

From a medical historian's perspective, the damage wrought by Queen Victoria's gene was grim: three affected children, six grandchildren, and seven great-grandchildren. Ten male sufferers and six known female carriers. Only late in life were her eyes opening to the depth of the devastation. "Our poor family," she wrote privately, "seems persecuted by this awful disease, the worst I know."

. . . .

AS DISEASE NAMES GO, I HAVE TO WONDER WHAT IN THE WORLD Dr. Johann Schönlein was thinking when he came up with *hemophilia* back in the early 1800s, a term translating from the Greek as "love of blood." I have it on good authority that affection has nothing to do with it. Aggravation is more like it, according to Christine Pullum, a dry-humored Louisianan who is one of those rarest of the rare, a woman with classic hemophilia. An acquaintance had introduced me to Christine, age sixty-six. I reached her by phone at her home in Lafayette, where she, a retired administrative assistant, lives with her husband of thirty-six years, Doyle, a former junior high school math teacher.

With so few cases of female hemophilia A on record, it would be a stretch to call any one instance typical. Still, experts say it most likely would manifest in a daughter born of a male hemophiliac and his carrier wife. Not so with Christine. "Our family didn't have the usual pattern," she told me. "I got it directly from my daddy." In a genetic fluke, the good X chromosome she got from her mother failed to compensate for the defective one, making Christine what's called a symptomatic carrier. Her two sisters are also carriers but have never been symptomatic. The shared disease between Christine and her father created a unique bond. Throughout her childhood, he would tell her that their hemophilia could be traced back to the czars of Russia, referring, I gathered, to the doomed Romanov family, all of whom—Nicholas, Alexandra, Olga, Tatiana, Marie, Anastasia, and Alexis—were assassinated in 1918 in the wake of the Russian Revolution. Christine has never seen any documentation of this tie, she acknowledged, and I had to admit I couldn't see how this branch fit onto the Romanov family tree. Even so, it sounded like a sweet way to make a daughter with a scary disease feel special.

To help build a picture of who she is now, I asked Christine how her friends would describe her, which, after a moment's embarrassment, led to an interesting cavalcade of disclosures. "Well, I'm the opposite of a hyper person," she told me, "and I'm going to be on the petite side in the dress department." She used to be five-three, she added, but has lost two inches to osteoporosis. Describing her ethnicity as "a half-breed," she chuckled and elaborated that her father was from Greece, her mother, American.

"My daddy knew when I was a year old that I was going to be a bleeder." One day she'd bitten down on the edge of a coffee table, something not unexpected in a teething child, and cut the inside of her mouth, a wound that was slow to stanch. This was Jackson, Mississippi, in the late 1930s, she recalled, a time when you couldn't convince a doctor that a girl had hemophilia and, even if you could, there wasn't a heck of a lot anyone could do about it. So her father, who was born in 1903 and lived to age seventy-five, taught her what would be the secret of his longevity—being very, very cautious. The first modern medical treatment for hemophilia didn't arrive until the early 1940s in the form of blood transfusions. These weren't a perfect solution. Receiving another person's whole blood or plasma did not provide what was most needed, a concentrated dose of clotting factors (not until the late 1960s was such a transfusion possible). But it was enough to save lives. With that in mind, Christine could look back with no nostalgia whatsoever at what her father had faced when he was small. "Daddy's treatment was spiderwebs," she marveled, "the soot from the chimney, things like that"—folk remedies used for crudely plugging breaks in the skin.

My telling Christine that I grew up with five sisters seemed to free up some thoughts on the bane of her teenage existence, menstruation. "Of course I had the very heavy periods, where I couldn't even go to school." This was back in the day when it was deemed inappropriate for women to wear pants, she explained. "We always wore skirts, and blood would come all the way through." Pants, at least, would've kept the problem a little more hidden, but, without that option, "I knew what I had to do: I stayed home." Her doctor was hardly a lighthouse in the storm. He suggested, she recalled, that heavy bleeding might simply be perfectly normal for women in her family. So, chin up, not to worry. Only after the birth control pill became available in 1960, when she was in her early twenties, did Christine experience a normal cycle. Oral contraceptives have since become standard protocol for many women with bleeding disorders.

Christine grew weary over the years of having to convince doctor after doctor that she had what was unanimously considered "a man's disease." A formal diagnosis of hemophilia didn't come until she was a married woman in her midthirties, following treatment at the University of North Carolina medical school. She'd returned home to Lafayette with a document quantifying her factor VIII deficiency and providing instructions to her local doctors—that, in short, when future bleeding problems occurred, she must be "treated in the same manner as a man with hemophilia A."

"It was a relief just to have it on paper," Christine told me, the memory of the milestone still making her voice dance. Unfortunately, the paper was put to almost immediate use. In 1975 she was hospitalized with severe hemorrhaging in her abdomen, an extreme example of the spontaneous and painful internal bleeding hemophiliacs can experience. Once again, though, doctors disbelieved her claim of hemophilia. Thankfully, Christine survived the time it took Doyle to race home to get "the proof."

The letter of diagnosis also addressed a more personal issue that Christine had broached during her workup, the risks of having a child. What for many is a joyful decision was for the Pullums a complicated one, requiring an emotional, medical, and financial accounting of where they stood. What were the health risks to Christine of carrying a child to term? What was her own independent prognosis? Could they accept the fifty—fifty chance of a son being a sufferer or a daughter being a carrier and, possibly, symptomatic like her? And what if an affected child had a form of hemophilia more severe than Christine's? Setting emotions aside, could they, bottom line, afford to care for such a child?

During Christine's hospitalization, the Pullums experienced the sticker shock of a top treatment of the time, cryoprecipitate, an expensive extract of clotting factors made from fresh frozen plasma. They were also mindful of the challenges faced by Christine's sister, whose young son had been diagnosed with severe hemophilia two years prior. Plus, for Christine, the idea of motherhood was tempered by her memories of childhood, how simple events such as losing baby teeth or falling off a bike were ordinary for everybody but her. She shuddered at the thought of her child going through what she'd faced as a girl.

Today a mother-to-be can undergo amniocentesis to learn if her baby has a genetic disorder such as hemophilia, but subtract three decades and, of course, the diagnostic tools weren't nearly as sophisticated. In Christine's original letter from her doctor at the University of North Carolina, he had explained in stark terms one of the few options available to her. "There is no test at present which will tell us during pregnancy whether a fetus does or does not have hemophilia. However, it is possible to determine the sex of the fetus" at about the fifteenth week, he noted. "Some couples in which the wife is a carrier choose to use this method to have only girl children by planning an induced abortion of any male fetus." In the final balance, Christine and Doyle chose not to have a child.

As our conversation spun back to the present day, Christine reflected that the intervening years have been largely free of serious complications from hemophilia. Her extreme vigilance has paid off in that she's rarely needed infusions of clotting factor. She is, however, of the generation for whom treatment came with disease. It took just one cryoprecipitate infusion in the mid-1970s for her to become infected with the hepatitis C virus (HCV). As she approaches her sixty-seventh year, Christine must contend with cirrhosis of the liver. She now has a diagnosis doctors will believe.

I doubt Christine would ever consider herself an activist but, in the next breath, she explained how she and Doyle founded and facilitate a twice-monthly HCV information exchange/support group. Now, this is a colorful group, she admitted, drawing individuals from the Lafayette area who have a disease in common but sometimes little else. In addition to folks with heart disease, diabetes, and bipolar disorder, "We have people who are co-infected with HCV, hepatitis B, and HIV. You think you have problems, you should talk to a person like that. That is really, really tough."

For peer support regarding her bleeding disorder, Christine turns to what is sadly becoming a lost art, letter writing. One longtime pen pal has been Cindy Neveu, the Bay Area woman who'd introduced me to Christine. Like members of the Pullums' group, Cindy is multiply diagnosed. She has HCV and HIV, yet these don't even top the list. And it's literally a list. Her multiple medical conditions far exceed the surface area of her MedicAlert bracelet, so, as Cindy showed me one day, flashing the bracelet with a mock QVC flourish, it details only her most immediately perilous condition: her fibrinogen deficiency, a blood disorder that's believed to affect just one in forty-three million people.

Cindy, with whom I'd spoken by phone several times, had invited me to her weekly cryoprecipitate infusion session. The hour-long procedure was just getting started when I arrived at the infusion center in Berkeley's Alta Bates Hospital at ten thirty on a Monday morning. I pulled up a chair as the nurse, Carrie, swabbed the skin just beneath Cindy's left collarbone, the area where your index fingertip would rest if you were saying the Pledge of Allegiance. Here was Cindy's venous port, a rubber device the size of a nickel implanted under her skin. Since her last infusion, the skin had barely had time to heal over, and now Carrie punctured it again. The infusion needle looked like an extra-large thumbtack. Blood immediately swirled up the attached tubing—a good sign, the port was still viable—and Carrie started a saline drip, which shooed Cindy's blood back into her body. A nurse materialized just then bearing a plastic pouch containing what looked like melted orange juice concentrate. "Ah, here it is," Cindy said, "here's my cryo." Within seconds, the bag was dangling from the IV pole and the infusion under way.

Sometimes the session is not so flawlessly choreographed. Kept frozen at the local blood bank where it's produced, the cryoprecipitate must first be slowly thawed before being couriered here to the hospital. Traffic snarls can hold things up. The window for delays is short, however; cryo loses its efficacy within four hours of thawing.

Carrie promised to check back in a few minutes, and Cindy settled back in the infusion bed, a kind of rose-colored La-Z-Boy in permanent recline position.

"I've been doing the cryo forever," Cindy immediately confided to me, putting a playful exasperation into her words. "I was diagnosed at birth with the fibrinogen deficiency when my umbilical cord wouldn't stop bleeding. The doctors put it together pretty quickly, though, since my older brother also had it."

Unlike hemophilia A or B, where inheritance is linked to the X chromosome, a fibrinogen deficiency is autosomal recessive, which means both her parents carried the faulty gene but neither Mom nor Dad had bleeding problems. In other words: "Prior to my brother and me, there was no family history."

Within the body, fibrinogen (also called factor I because it was the first of the thirteen clotting factors to be discovered) is the last step in the coagulation cascade, the "glue" that holds a clot together. By contrast to people with hemophilia, Cindy explained, those with a factor I deficiency are more likely to "have bleeds" in muscle tissue and the mucous membranes than in the joints. Treatment options also differ, which is something of a bone of contention for the thirty-six-year-old. A hemophiliac today can get a prescription for cutting-edge formulations of factor VIII or IX—genetically engineered, not distilled from human blood. These come in a powdered concentrate that a person simply reconstitutes and injects. Cryo, by comparison, is made from a decades-old recipe: You freeze a healthy donor's plasma, then defrost it. The solid material that collects at the bottom is the cryoprecipitate, rich in all the clotting factors, including the fibrinogen someone like Cindy needs. From this point on, the blood factors aren't further separated. So to raise her fibrinogen levels sufficiently, Cindy needs to infuse a bag containing five donors' cryo. Her brother, Dave, has to receive five times that amount on a weekly basis.

The transfusion isn't cold, Cindy explained. It's not the body slam of chemotherapy or the head rush of too much caffeine. Nor does it hurt, which made sense once she'd said it. This was *replacement* therapy, after all—simply adding something the blood is missing. But there are risks. "With any blood product you can always have a reaction," Carrie had told me earlier as she checked Cindy's blood pressure and temperature. While donated blood is now elaborately screened and tested for hepatitis and HIV, for instance, sometimes a bug or bacterium can slip through. Unlike other blood products, cryo cannot be heat-treated or "washed," to borrow the nurse's word. Though a safe, genetically engineered factor I concentrate is manufactured overseas, it has not been approved by the FDA and therefore cannot be legally obtained in this country.

"I would love to have the concentrate," Cindy mused, easily envisioning how it would simplify her life. "I could do it at home. I could treat prophylactically much more efficiently. I could travel!" As she sees it, however, the reality is that the profit potential just isn't great enough for an American pharmaceutical company to create a comparable product. There may be as few as seven "factor I's" in the United States, according to one statistic. Hence cryo, the standard treatment, which, as Cindy wryly put it, is "cutting edge, 1972."

I can't imagine anyone more deserving of greater simplicity in her life than Cindy. But it may not be in the cards. On top of the fibrinogen deficiency, she has an unrelated clotting disorder. This means that, while she has to infuse once weekly to, in essence, thicken her blood, she must also take medicine twice daily to thin it. In addition, and unrelated to these blood disorders or to HIV or HCV, she has a rare neurological disease, transverse myelitis (TM), a condition similar to multiple sclerosis. I don't think there's a satisfying expletive that sums up having five life-threatening illnesses.

"My system doesn't know if it's coming or going some days," Cindy added. In fact, her medical history is so complicated that she carries with her at all times a single document that spells out the entirety of her conditions. "Here, I'll show you." She pulled her wallet from her purse and removed a fold of paper. Regardless of the emergency, she noted, "The first thing a provider wants to know is, who's going to pay for this? So that's at the top." Underneath this insurance info was a summary of her medical history, single-spaced. On the flip side: the names and numbers of her various doctors as well as a listing of her many meds. "The day you don't feel good is not the day you want to be explaining all this."

Cindy also carries a MedicAlert card—call the number and you'll get her health history over the phone in a choice of languages. "And for airports, I've got this," another ATM-like card. "It says that I'm bionic."

"*Bionic*? You mean, from your brace?" A metal brace provided support for her lower right leg, weakened by the TM.

No, Cindy answered with a laugh. To control leg spasms, she had to have a catheter implanted in her spine to deliver antispasmodic medicine automatically. "Very cool! But now I beep in metal detectors."

Next, Cindy handed me a letter from her doctor, its edges showing signs of frequent use. Unlike the densely typed medical history, this page consisted of a single brief paragraph. "To whom it may concern," it read. "Miss Neveu is very knowledgeable about her symptoms and the care required to treat her disorder." In case of an emergency, "PLEASE LISTEN TO THIS PATIENT."

"That sure says a lot about the people you have to deal with."

"Yeah, this is good to have," she said. "Nobody believes me when I start telling them what I've got." Her light blue eyes sparkled. "I'll flash it now and then, and say, 'Do not mess with me.' "

Cindy seemed to be drawing to a close her show-and-tell when I, like a nosy brother, prompted, "So, what else you got in there?" Her eyebrows lifted as she grabbed her purse: *What indeed?* She then rummaged and pulled out a cell phone—"Only for emergencies," she explained. The phone company offers an inexpensive monthly plan for disabled people. Next out, an antique cough drop and a floaty pen, in which a BART train floats between San Francisco and her hometown, Oakland, followed by several old appointment cards for doctor visits—Cindy sometimes has six appointments a week. "You get really nasty hands driving a wheelchair," she confided, now flashing a packet of Wet Ones. It was only then that I noted the wheelchair folded up and stowed to her left. Though she can walk with a cane, the wheels provide a speedier transport. One thing conspicuously absent from the purse of a woman Cindy's age, though, was a tampon or two. Well, she explained, it's too dangerous for her to bleed, so she takes an "industrial-strength" birth control pill that keeps her from ever having a period. This has been necessary since puberty.

The final stop on the pocketbook tour was a small case, imprinted with GIRLS RULE!, that contained business cards for her Web site, the sublimely named Shemophilia.org. Cindy explained that a former hemophilia treatment nurse here at the hospital had affectionately called all the female patients shemophiliacs. As Cindy was preparing to launch her site in 1999, the nurse graciously allowed her to, as Cindy put it, steal the name. Already in existence were informational sites focused on some of the more prevalent bleeding disorders, such as von Willebrand's, but Cindy felt that "too many factor I's and II's were falling between the cracks." The potential for feeling disconnected is far likelier when you're just one of a handful of women in the entire country with, say, a fibrinogen deficiency or classic hemophilia. What was needed was an online "community center" for these individuals. Cindy hoped the site would become a forum where someone such as Christine Pullum could share her wisdom.

With the playful name Shemophilia, Cindy also sought to convey the user-friendliness rarely found in a pamphlet picked up at a doctor's office. I think her instincts were spot on. FibrinogenDeficiency.org just wouldn't have the same zing. Among people with bleeding disorders, she added, the term *hemophilia* is a kind of collective shorthand. In fact, she said with a laugh, when she's tried to explain to someone her specific disorder, their eyes have tended to glaze over. "It's often just easier to tell people I have hemophilia."

In my experience, I've found that people who are strong advocates for peer support either never enjoyed it themselves or are taking the opportunity to pay it forward. Cindy is clearly in category two. In December 1991, two years after learning she'd been infected with the AIDS virus from tainted cryoprecipitate, she arrived at an HIV support group in nearby Pleasanton. As she took a seat in the host's living room, she realized she was the only woman in a small circle of gay men. "At first, I was kind of afraid because I'd lived such a sheltered life and didn't have a lot of experience with the gay world," she recalled. "But from the start, they were so welcoming and wonderful. I called them My Tuesday Guys, 'cause that's when we'd meet. And we'd always go out afterward—this chick with a cane and eight really good-looking gay men going out to Bakers Square." She hooted with delight at the memory. "I'm sure people had to wonder."

Much of her present-day passion for mentoring was inspired by the man who'd started the support group, Wes. "He taught me a lot of important things—about treatments and disability benefits and working the system." Wes, who passed away five years ago, was also a buoy during some rocky emotional times.

In the same spirit, she's now mentoring a fourteen-year-old fellow factor I in Ohio, keeping in touch through e-mail and phone calls. Further, Cindy sees it as part of her role with the Web site to facilitate a mentor matchmaking of sorts. "Usually I'm able to link people pretty well. If you contact me and say, 'I have factor VII deficiency, I want to talk to somebody,' I'll make a couple of calls and get you connected with another woman who has your disorder." She'll make sure, too, that you receive written information and, if need be, a referral to an accredited hemophilia treatment center.

"That's the biggest problem with the rare disorders," she continued. They're often not seen as falling under the umbrella of hemophilia, so these patients don't get referred to the best treatment. Sad but true, "There are all these independent docs out there winging it on old, old information." Cindy told me about a fifty-year-old factor I with whom she'd spoken not long ago. "She was only being treated with plasma, not cryo—very inefficient, because you're getting a lot less of the clotting factor and a lot more volume." Though Cindy has grumbled about how dated her cryo infusions are, this woman's care was cutting edge, 1940.

"That's like going to your AIDS doctor and only getting a prescription for AZT instead of a multidrug cocktail," I observed.

"Yeah, it's pretty scary."

The amount of misinformation is sometimes daunting, Cindy acknowledged. Not long ago, for instance, she had to swoop in and set straight a nice woman from Maine who, for reasons unknown, believed that bleeds "were only serious if they occurred below the knees!" Cindy actually went and met this woman and her husband in person. She arrived at the restaurant they'd chosen to witness the husband rearranging all the furniture in the waiting area, so that his wife's knees would not be imperiled.

At this point, although Cindy's stories hadn't run out, her cryo had. Carrie returned and removed the empty pouch as Cindy, like the patient schoolteacher she used to be, provided a careful narration of what followed: "Now she's going to flush the line with saline to clear it, then put in a little heparin—that's an anticoagulant—to keep the port from getting clotted." Carrie then unplugged the plastic tubing, bandaged the tiny hole, and Cindy was free. "If everything goes right, if I don't fall or something," she said matter-of-factly, "I won't need to come back here till next Monday." And with that, we said quick goodbyes. It was 11:30 A.M. and Cindy Neveu had much more to fit into her superhuman day.

ELEVEN

Blood Drive

IN A BODY AT REST, A SINGLE BLOOD CELL COMPLETES A full circuit of the circulatory system in just about thirty seconds. Blood bursts from the heart at its top speed, around one mile per hour, and shoots through the tough plumbing of the arteries outward to the body's extremes. On its return, venous blood—now depleted and slogging waste—often must work against gravity and, at best, will only reach half its starting speed. In other words, the second half of the trip is more arduous than the first, which I suppose could be said of life as well.

At an approximate rate of one million per year, my blood's clocked just over forty-three million circuits and, barring catastrophe, I anticipate another forty million or so. Long-livedness runs in my family, though I can expect the expected "infirmities of old age," as my eighty-seven-year-old great-grandma Bridget's obituary described her last years. Already though, as all people do, I've outlived my blood many times over. I contain, for instance, no red cell older than four months, no platelet over ten days. Some of my white cells survive less than six hours. Other blood cells are longer-lived, it is true. The lymphocytes called memory cells, for example, are ferried about the circulatory system for decades. But these, too, eventually fail. Such deaths go unnoticed by me but not by my body. Individual blood cells are constantly being replenished or replaced through a remarkable system of inner housekeeping. That being said, it is still a disquieting notion that my blood retains none of its original parts.

Once blood is removed from the body, the cellular life spans plummet. Forestalling the death of blood is the major clinical function of any modern blood bank, although, granted, you wouldn't find such dark phrasing in an annual report. A blood bank gives the impression of being less a bank—as in, a place to stockpile—and more a hospital, wherein blood is on continual life support. I quickly came upon this realization during my tour of the main branch of Blood Centers of the Pacific, a state-of-the-art facility here in San Francisco, where I'd come to see how blood products are made. As Richard Harveston, the director of hospital services and my genial host, explained to me, blood must be carefully housed, nourished, and tended. "Blood is a living tissue," he said. And therein lies the challenge.

"One of the greatest advances in blood banking came in the early 1970s with the advent of plastics," continued Richard, in what at first seemed like narration from a different tour.

"Plastics?"

"Yep, just like that guy says in the movie *The Graduate:* 'Plastics.' " He smiled, adding, "Before technology for making plastic bags was perfected, blood was drawn in glass bottles," which caused a lot of headaches. Bottles took up lots of storage space and trapped air, fostering bacterial contamination. By contrast, plastic bags provided a slew of advantages, including being virtually unbreakable, lightweight, airtight, and malleable. In addition, Richard explained, "Plastics made possible the era of component therapy." This last sentence rose to a deliberate crescendo. In component therapy, blood is separated into its various parts, which then become highly efficient, targeted interventions, such as the present-day replacement therapies for hemophilia. Cindy Neveu's cryoprecipitate, although a dinosaur when compared to other treatments, would also fall under this heading.

We were standing on the periphery of the blood center's collection area, where five donors were giving blood. To illustrate the procedure, Richard pulled out a "blood collection set"— three connected clear plastic bags (a large and two smalls) trailing a tangle of tubing. The whole mess looked like a jellyfish, the kind of thing a kid on a beach would poke with a stick. "Blood flows into here," he said, pointing to the primary collection bag. This pouch already contained a small mix of fluids: an anticoagulant, a phosphate to maintain the pH, and a nutrient to keep the blood cells alive. He then traced the tubing to the second bag, which would later be used during the blood processing phase to hold the plasma, and to the third, a pouch for platelets. "If you'll notice here," he said, inviting me to feel the last pouch. "This is a different plastic, this has a different porosity," allowing gases to pass in and out. "Just like we do, platelets have to breathe."

Leaving behind the homey atmosphere of the collection area, Richard and I entered the factory-like environment of the Component Lab and stepped around what could've been a beverage cart from an airplane, heaped with fresh pints of whole blood. It looked kind of disorderly, truth be told, but each unit, Richard hastened to point out, was bar-coded, its every movement through the facility tracked. For each bag here, a tubette of the donor's blood had also been collected and affixed with a matching bar code. These samples were already on their way to a lab in Arizona, where each would be comprehensively tested for HIV, hepatitis, syphilis, and so on, and so forth.

Within six hours of being drawn, a large portion of the pouches are spun in a centrifuge. The technician working this machine allowed Richard to demonstrate. The centrifuge, whose interior is chilled to just above freezing, has six pewter buckets. He stuffed each with a bag of blood, the one full and two empty pouches sandwiching nicely. Blood naturally separates, Richard noted, but this device speeds up the process. "Once it gets started, the buckets spin out, like being in a Tilt-A-Whirl." The speed of the "ride" can be varied, he added. A light spin, for instance, is necessary if you're harvesting platelets. Richard then closed the lid and flipped a switch. The procedure would only take a few minutes.

In the interim, he steered me to an adjoining room, a kind of platelet pantry. On floor-to-ceiling shelves, small bags of the straw-colored cells lay like flat pillows on undulating metal racks, the rocking movement driven by grinding motors. "Platelets are very fragile," he said in a raised voice, "not hale and hearty like red cells." But they're very eager to clump, which is their pivotal role in the coagulation cascade. Once clumped, though, they don't unclump. "So," Richard concluded, "you have to keep them in constant motion." You also have to keep them exactly at room temperature, a fact I found quite odd. Ironically, once removed from the 98.6 degrees Fahrenheit of the human body, platelets no longer thrive at that temperature. He picked up one of the bags for me to peer at, holding it up to the ceiling light. The platelets were just a swirl in a shallow bath of plasma.

These cells remain functional for only five days, Richard said. They're the shortest-lived products produced at the center, and biting into that time is the thirty-six-hour wait for test results. I quickly did the math in my head. Subtract the day it takes to process the blood, minus the day and a half for testing: "So half their shelf life is spent here on the shelf," I observed.

"You got it," Richard said with a nod. "Which is why we always, always need new donors."

Red cells, he went on to say, can last forty-two days if refrigerated and years if frozen. Plasma is more finicky. If not frozen within six hours, the essential clotting factors "will disintegrate" or break down. Frozen plasma will keep for no more than twelve months.

Back at the centrifuge, Richard gently withdrew a spun bag of blood, now displaying neat layers of amber, white, and burgundy. We took a giant step to the right, at the same time moving from high tech to low. At this workstation each bag of separated blood is hung by its edges to the "plasma expresser." Anyone who's worked an old-fashioned orange juicer could handle this device. Pulling down a simple lever applies pressure to the lower portion of the bag, thus squeezing the plasma at the top through the tubing and into the second collection bag. The only trick is knowing when to stop pulling.

On the other side of this counter, an IV stand held several fat red pouches of plasma-less blood that were now being stripped of white cells, a slow process that appeared to depend mostly on gravity. Blood snaked down a length of narrow tubing, passed through a white-cell-catching filter about the size of an ant trap, and pooled in a bag near the floor. The white cells would be discarded. Watching this procedure brought up a question that's nagged at me for many years: If a healthy person's immunity is largely contained in his or her white cells, couldn't an ill person benefit from them? Or, coming at it another way, why throw them out? Wouldn't transfusing them be useful?

"No, almost never," Richard answered. "White cells are not a good thing, and you want to remove them." Beneath his blanket statement were a number of powerful reasons. For starters, too great a risk exists of transmitting an infectious disease for which testing isn't done, such as the cytomegalovirus (CMV), which may be present in white cells even if the donor has never manifested symptoms. Further, contrary to my layman's thinking, white cells rarely see another person's white cells as allies. Instead, they go on the attack. The recipient may, as a result, suffer a high fever or a life-threatening reaction. "There are very few indications for white cell transfusions," Richard concluded, "one or two a year, if that." Thankfully, for the vast majority of patients a far safer and more effective alternative exists in antibiotics.

Okay, that all made sense, but a new question now replaced the old. Of course, I began, people with HIV cannot and should not give blood. But in view of the fact that (1) HIV only infects white cells, and (2) white cells are removed from all donations, why then, speaking hypothetically, couldn't a person with HIV give blood?

Richard's whole demeanor said, *Ah, good question!* "Well, you're right, HIV does only infect white cells," he replied. But when you're "manipulating" blood products, it's not so cut-and-dried. He then motioned to the centrifuge. "Centrifugation is a pretty crude separation technique. You're artificially causing trauma. White cells are very fragile; they can rupture and release the virus." Free virus, he called it, which can then turn up in the wrong blood "zone." In test studies with HIV-infected blood, he noted, "All of the blood products have been shown to contain virus"—red cells and platelets, as well as plasma.

As for my hypothetical, as it turns out, people with HIV do on rare occasion donate blood, although they're typically unaware at the time of their positive status. "We get about two or three HIV positives a year out of 125,000 donations, which means the donor history and medical screening we're conducting is actually quite effective."

By way of explaining to me the many layers of safeguards in place, he suggested we jump ahead thirty-six hours in the processing of newly donated blood. At this point the Arizona test results have just arrived by computer. The red cells have remained refrigerated, the plasma kept frozen, and the platelets have never stopped undulating in their metal beds. Richard and I now stood in the Label & Release room, where a technician sat before a computer monitor, a box to her right filled with rock-hard plasma units. The technician swiped the bar code on the first frosty unit of plasma, calling up its results, pass or fail. (The red cells and platelets would also undergo this inspection today.) An A-OK was followed by a search of national, state, and internal databases to quadruple-check the donor's information. Has the individual ever been deferred in the past because of, say, foreign travel or short-term illness? Did the donor make his or her donation before the mandatory fifty-six-day wait? (This waiting period allows hemoglobin levels to return to normal.) An approval label only generates if no flags go up. The rejection rate is "very low," Richard noted. "Far less than 1 percent."

The new label was smoothed over the old, then the unit was scanned again and officially validated. "Now this unit is, by definition, 'Finished Goods,' " Richard announced, framing it with his hands.

Along with the seal of approval, each bag at this moment acquires an important quality: monetary value. The not-for-profit Blood Centers of the Pacific does, after all, have to survive financially. In the Bay Area this translates into a unit of fresh frozen plasma selling for $70; red cells for $180; and platelets for $600. (The national average price tags are roughly 20 percent lower, Richard noted.) Yearly, the center sells to forty local hospitals approximately 125,000 units of red cells, 50,000 of plasma, and 15,000 of platelets. The organization wholesales an additional 75,000 units of plasma to pharmaceutical companies for further processing, such as the making of factor VIII concentrate. (Only a small percentage of the center's total output is whole blood, a fact I found surprising. TV medical dramas, as it turns out, vastly overplay the call for whole-blood transfusions.) At any given time about 10,000 units of special red cells remain here on the premises, 99 percent of which are kept frozen in long-term storage. In this capacity, the bank is most bank-like. Some of this store is autologous blood—donated and reserved for an individual's own future use, such as for an upcoming surgery—but most of it is blood of the rarest types, the Château Lafittes of the blood world. To keep these red cells viable for as long as possible, each unit is infused with a preservative, Richard explained, "not unlike the antifreeze you put in your car." The blood is not left in the bag but spread flat in "a single monolayer," he continued, and frozen at minus eighty degrees Celsius. *That,* I thought, *should be interesting to see.*

But first he led a winding route from the Label & Release room to a set of windows overlooking another work space. The sight of two white-coated scientists huddled over computers wasn't all that fascinating, but the work these gentlemen were doing certainly was, Richard assured me. He gestured grandly. "This is our Immuno-Hematology Reference Laboratory. It is very much a part of the history of blood banking, one of the oldest in the country, and one of the most famous." Perhaps he could tell by the look on my face that I'd gotten lost on the way through his hyperbole. Richard paused, then rewound his narration. "You know, each individual has a 'genetic fingerprint,' if you will, on his red cells—"

"Right. Not a DNA signature, but a kind of tag that identifies your blood group."

"Yes," he nodded. "And the most significant in transfusion therapy are the best known—A, B, AB, and O."

Sure, anyone who's donated blood knows these letters. And of course I couldn't help noticing them prominently displayed on every blood product. This hematological safety code devised in 1901 put an end to hundreds of years of dangerous blood transfusions, I knew. As is often the case with scientific breakthroughs of this sort, the discovery of blood types began unceremoniously, with a curious individual trying to unknot a puzzle. Austrian pathologist Karl Landsteiner could not fathom why adding a bit of one person's blood into test tubes of other people's blood caused such varying results. Sometimes the red cells bunched together, sometimes they burst, and sometimes there was no reaction at all. Now, this was not an unknown phenomenon. Earlier scientists had concluded that these cellular dynamics were due to a clash between healthy and sick blood. Landsteiner, however, was using only the blood of healthy subjects, including his own. With the kind of glee I imagine only the fussiest scientists having, Landsteiner pulled out his graph paper. He mixed and mixed and mixed, taking careful notes and charting his findings. Patterns emerged, and he identified three groupings of blood—blood groups—which he labeled A, B, and C. (C later became O.) As it turned out, Landsteiner belonged to this last group, type O, making him what is now called a universal donor. In terms of his experiment, this meant that his red cells didn't react to any other specimens, which, in an odd way, is the aspect of his story I enjoy most. Even his cells, it seems, were dispassionate observers.

In a more technical sense, what Landsteiner had documented was a classic antibody-antigen response. For the purpose of illumination, consider a patient with type A blood (which means his red cells have the A antigen). If wrongly infused with type B blood, his body will immediately launch an assault: his antibodies versus the foreign red cells. (A similar scenario will play out if a type B patient is given type A blood, or if a type O patient is transfused with either A or B blood.) It's not surprising that, in his initial experiment, Landsteiner missed the fourth blood type, AB, which, for example, is found in only 4 percent of the U.S. population. These individuals are called universal recipients because they can safely receive any blood type.

One last major antigen is noteworthy here, antigen D, whose presence or absence is indicated with an Rh+ ("Rhesus-positive") or an Rh−. If an Rh− woman is carrying an Rh+ child, the mother-to-be may produce antibodies that will threaten the baby's life, a condition that, fortunately, can be identified and treated. Karl Landsteiner co-discovered the Rh blood factor in 1940, ten years after winning the Nobel Prize for his ABO blood grouping system.

Richard cleared his throat. In addition to the A's, B's, and D's, he explained, "There are literally hundreds of other antigens and proteins—both on the red cell surface and embedded in the red cell membrane—that are 'genetically informed.'" Normally, these don't figure prominently when a person needs blood. Some people are born with unusual blood, however, and others develop it. For instance, patients who have received multiple transfusions may have accrued antibodies to these minor antigens. Over time these individuals become harder and harder to match with suitable blood. That's where this specialized lab comes in. As Richard explained, the researchers obtain samples of blood from hospitals throughout Northern California and test and catalog these unusual antibodies, antigens, and proteins. The lab also houses a national rare donor blood registry and provides a round-the-clock consultation service for hospitals trying to hunt down rare blood.

"The thing about rare blood is, it's *rare,* so people don't need it very often. But when they need it, they need it *now!*" By means of demonstration, we stepped into a nearby room with a small rumbling fridge. Richard randomly pulled out a pouch of red cells and read off for me a sampling of its unusual markers: "So this unit is big C positive, little e negative, little c negative, big E—these are all different antigens, these are Rh—and this is Kell . . . big K and little k; Duffy a, Duffy b; Jk(a) and (b), M, N, S, Lutheran, and Kinney!" Had I not glanced over the list, I'd have thought he'd made up some of those last ones. But no, this specific blend will likely be called for today, he predicted, and within hours will be coursing through someone's bloodstream.

While at any given moment a small number of fresh units are available, the bulk of the center's inventory is frozen. A quick walk took us to the deep freeze, a dim room dominated by eight coffin-style industrial freezers. You could probably fit several bodies into just one, I figured, and in fact the eleven hundred units stored within each do add up to about twenty quarts of blood, or four bodies' worth. Richard creaked open the hood of the closest vault. Iced-over metal containers the size of clipboards were arranged like hanging files. Inside these, he explained, the cells are a thin red crust. When a hospital requisitions a unit, "It is thawed in a seawater bath at body temperature"—which struck me as a lovely way to emerge from so harsh a hibernation. Next, the "antifreeze" is removed, leaving the red cells ready to be shipped and transfused. The sell-by date for deep-frozen blood has yet to be established, Richard noted. Ten years is the industry guesstimate, but, he ventured, "It probably is almost good forever."

As Richard steered me back to where the tour had begun, I marveled at the support structure in place here within this enormous stretch of a building: a staff of 350, a thirty-million-dollar annual budget, a steady hum of technology, all devoted to sustaining these small bags of fluid and readying them for their eventual return to circulation. While I'm sure it wasn't Richard's intent, my witnessing the effort expended on blood's behalf actually left me more in awe of what the center tries so hard to replicate, the perfect packaging of the human body.

Back in the collection area, one such specimen sat within the contour cushions of the blood center's latest high-tech toy, the e-chair. Richard said in the quiet voice of golf commentary, "This is the wave of the future." He gently pulled me off to the side so we wouldn't be hovering too close to the young Latino donor. The e-chair, Richard explained, is a six-months-new machine that, in one sleek apparatus, performs all the tasks of the Component Lab. It does this by extending the circulatory system by a few feet. The donor's blood passes through tubing that disappears into a chairside contraption, about the size of a two-drawer filing cabinet. Inside is a whirring centrifuge. The wonder of this machine, Richard said, is that you can program its computer to harvest only the specific cells you need. The unneeded blood is then neatly transfused back into the body through a second tube. The technical name for this process is *apheresis,* which sounded to my ear like a condition requiring anti-itch cream but, in fact, comes from the Greek for "to separate."

To underscore the machine's efficiency, Richard had me imagine the typical pouch of whole blood that's collected in a standard donation, from which only a tiny squirt of platelets is derived. In a single apheresis session, he stressed, five times the number of platelets can be withdrawn without removing a single red cell.

One aspect of the e-chair setup that's consistent with the traditional method is that the blood collection bags are not in the donor's direct line of sight. This, I am certain, is for the best. No matter how soothing the phlebotomist's manner or how frequently the donor donates, I don't believe a person could stare directly at that transparent pouch as he or she drains into it without having a visceral reaction. I suppose this could be considered the most primal form of separation anxiety: a person from his or her own blood.

At this moment the young donor was being separated from both platelets and plasma. Richard pointed out the two corresponding pouches hanging behind his shoulders. The technology is cost-efficient in so many ways, Richard enthused, allowing his inner accountant to romp—lower labor costs, less lab time, fewer blood tests, the ability to maximize the contribution of a single donor. It's also safer for patients who receive the products. He used the example of someone being treated for severe leukemia. Such a person would likely need an infusion of platelets every other day for five months. As I'd already learned, one infusion is typically a pooling of five people's platelets. With apheresis, however, the product comes from just one donor. Thus the potential for a bad transfusion reaction is slashed 80 percent.

The only drawback, he conceded, is the time commitment required of the donor—ninety minutes as opposed to the usual half hour it takes to give a pint of blood. That's where the *e* in *e-chair* comes in. A mounted computer screen and keyboard let you access the Internet, listen to CDs, or watch TV or movies. A five-foot-tall, fully stocked DVD carousel stood off to the side. Conceivably, you can catch up on office work and e-mail while giving blood (though it helps if you're a good one-handed typist, since one arm is locked down with tubing).

The current donor, wearing earphones, was about an hour into watching the movie *X-Men* and looked completely absorbed. Of course, I wouldn't have wanted to disturb him—Mystique was just about to sabotage the mutant-detecting device Cerebro—but I was curious to know his story. One of the technicians had mentioned to us that he was a first-time donor, and Richard had quietly noted that he was much younger than their typical volunteer. In order to earn a seat in this e-chair, I knew, he'd had to pass an extensive donor screening centered on a medical history survey with forty-three yes-or-no questions. In the strictest sense, these questions are designed to eliminate groups of people. The first fifteen, asked during a brief one-on-one interview, are intimate in nature but clinical in wording. The questionnaire is also ever-evolving. As there is not yet a blood test for the human variant of mad cow disease, for example, questions regarding past travel to the United Kingdom were added to disqualify the possibly exposed. However, there's no space provided for elaboration—no room for the *Well, yes, but*s, as in, *Yes, I did stay in England for a summer, but I'm a strict vegan, so beef never crossed my lips. . . .* The screening doesn't bend. Nor does it address a broader field of inquiry, the prospective donor's character and intent, which, granted, has no bearing on the quality of the blood itself and, in any case, would be hard to assess through simple yeses or nos. Even so, this eliminates the first question I would ask of a person, one that's far too open-ended to be practical but whose answer I'd still like to hear: *So tell me, why do you want to give your blood?*

SOMEONE NEEDS HELP, YOU DO WHAT YOU CAN. CALL 911. STOP AT the scene of an accident. Or help chase the oranges when an old lady's grocery bag breaks. Any good person would hardly think twice. Granted, because of where I reside (San Francisco, on a spider's throne of fault lines); where I live (off a busy intersection without traffic lights, the site of countless fender benders); and where I work (at home, alone, which involves a fair amount of staring out the window, daydreaming), maybe I do give this more thought and put the sentiment into practice a tad more often than most. (*Too* eagerly, too, a certain neighbor could claim, as when I reported a fire one lovely evening last summer upon seeing flames leaping from behind the backyard fence of his building across the way. Four fire trucks converged, the building was evacuated, the street blocked off. I stepped outside into the scene I'd tipped into motion and felt a tug of war between *God, I hope it's not a bad fire* and *God, I hope it is.* Not a lick of flame was visible from out here, so the tenants milling about wore looks of inconvenience or irritation rather than fear. I stood next to a woman in slippers cradling a bowl of goldfish when the word came: "It was just a barbecue, and some dumb stupid idiot in love with his lighter fluid." I appreciated the fireman's redundancy as much as his comment once he'd learned I'd made the call: "You did exactly the right thing.") So in any event, when a call went out for Bay Area blood donors a few years ago—*Urgent Need! Critical Shortage!*—on TV news and in the papers, I thought, *Yeah, sure, I'll give blood. I'm perfectly healthy. Needles don't scare me. Blood, either.*

When I was growing up, a citizen's duty to donate blood was instilled at home, school, and church. It was a patriotic gesture, like voting, only young people could also do it. That giving blood was synonymous with citizenship was a message that even made its way into one of the comic books I read as a kid. In *Action Comics* #403 (August 1971), the good people of Metropolis were asked to participate in a one-of-a-kind blood drive. Calling all able-bodied volunteers: Superman needs a blood transfusion! Reading it today, I have to wonder if I paid attention at all back then to the science of the story, if you can even call it science. You see, Superman is infected with a killer microorganism, a sentient menace that has set up camp in his bloodstream. With just two days to live, the Man of Steel must find a way to outconnive the conniving "Micro-Murderer." The only solution, doctors determine, is to flush the microbe from Superman's body using hundreds of gallons of blood. The comic-book doctors call this a transfusion, but it's really more a circulatory system colonic. Superman makes a televised plea for blood, and the citizens of Metropolis can't move fast enough to go donate. Having been saved countless times over by the alien Man of Steel, these everyday people are overjoyed to have a tangible way to give back.

To be honest, I'd have recalled little of this story had Steve not dug up a copy for me. The cover, on the other hand, was far more familiar, not for the striking image of Superman lying unconscious on a gurney but for the fact that the third person in line to donate blood is singer David Crosby circa the 1970 *Déjà Vu* album, or, at the very least, a remarkable look-alike. In front of David is a boy who looks the age I'd have been when the comic was brand new, though the kid's on crutches and has a

*"Calling all able-bodied volunteers:
Superman needs a blood transfusion!"*
("Action Comics" #403 © 1971 DC Comics.
All Rights Reserved. Used with Permission.)

leg cast so I can't help but wonder how he won the race to be first in line. For the sake of a single dramatic cover image, the artist shows the transfusion already under way. The collected blood, in a huge IV hanging over the hero's head, is the same Superman-red as his boots, briefs, cape, and the bright S on his chest.

I joined the ranks of real-world blood donors as soon as I was age-eligible, sixteen. In fact, in my wallet I still carry my original Spokane Blood Bank donor card, Type A, Rh+, on the back of which my record of regular "deposits" throughout high school and college remains legible in indelible ink. Nowadays, though, I only keep the card on me should I ever need blood in a medical emergency, not because I regularly give it. The last time I tried was in 1984 during an employee blood drive at a large Seattle company. Together with co-workers, I lined up outside the humming bloodmobile in the parking lot and was at last ushered inside. As I sat to fill out the standard form, my eye fell upon a question I'd not seen before, one directed at men: "Have you ever gone down on a guy?" Well, that may not have been the exact wording, but the query's intent was clear. *Why, yes, I have,* I thought, with none of the nascent pride that, in a more intimate situation, might have accompanied this declaration. (I wasn't out to my boss, who was capable of scaring me into stunned silence first thing Monday mornings with his "So, didja get any snatch this weekend?") Perplexed at first, I quickly gathered that the questionnaire was fishing for gay men who might have AIDS, which seemed sensible, given the infections transmitted through blood products and the lack of a blood test to prevent such accidents. Still, I felt unaccountably ashamed of myself. Suddenly squeamish at the sight of blood, or so I claimed, I handed back the form, covered my unbandaged arm, and slipped back into the office. A purloined doughnut provided the phony proof that I'd gone through with it.

Fifteen years later, during the Bay Area's 1999 blood drought, I looked forward to rolling up my sleeve again and this time doing the deed. Since all blood banks had begun using the ELISA-HIV test (which detects antibodies to the virus) shortly after it was introduced in 1985, I assumed that the gay restriction had been eased. I myself had tested for HIV half a dozen times since then and had always come up negative. Twice shy, though, I went online to scan the donor guidelines before making the trip to the blood bank. Good thing. The restriction hadn't changed. It was easy to imagine the red rush of embarrassment at being told in person, "No, sir, you do not qualify."

The question in question, number 9, hasn't changed in wording since then, a fact I confirmed on my recent tour of San Francisco's blood center. Under current FDA rules, all potential male blood donors must be asked verbally during the screening interview if they've had sex, "even once," with another man since 1977, the year identified as the start of America's AIDS epidemic. If the answer is yes, regardless of whether the sex was safe or the partner HIV negative, he is barred for life from donating blood (*permanently deferred* is the official term). It so happens that my first homosexual encounter was in 1977 at age sixteen; even if I'd sworn off men right then, I still couldn't give blood today. To be considered a qualified donor, a healthy gay man needs to have been celibate for the past twenty-seven years, a prerequisite that leaves me pondering: Can a guy who's not had sex in over a quarter century rightly be called "healthy"?

For the sake of argument, I can set aside the fact that the majority of gay men are HIV negative and committed to remaining that way. I can also ignore the assertion that, should gays be welcomed, blood centers will be misused as HIV testing sites, the presumption being that, illogically, fearful individuals would prefer a test site that demands documentation of who you are and is answerable to the federal government. At the same time, I fully accept that gay men in general are considered a high-risk donor pool. But why is there such inconsistency between what's required of gay donors and other groups? Straight men who've had sex with a prostitute, for instance, are barred from giving blood for just twelve months after that encounter. The FDA, I've learned, has argued that a data deficit is what keeps them from giving gay men this same "temporary deferral." The agency simply has no solid statistics on HIV infection rates among gay men who've abstained from sex for a year or more, a statement that begs the question, Do they really have comparable data on straight johns? The yes-or-no format of the donor questionnaire is also problematic. It doesn't elicit an elaboration of an individual's history of unsafe sex or multiple partners, which many public health experts consider a more effective means of determining genuine risk. Under current criteria, a woman who's had unprotected anal intercourse with numerous partners of unknown HIV status could technically donate blood (though obviously such a person shouldn't do so) while a young, HIV-negative gay man who's had nothing but safe sex could not.

Of course, quizzing people about their sexual histories isn't foolproof. Having worked in AIDS education, I know that people don't necessarily tell the truth about their sexual past or may genuinely not realize if they've put themselves at risk. And with practices such as unprotected oral sex, there isn't consensus about what is safe. Ultimately, experts agree, the best test is blood testing itself. That being said, the ELISA, an effective test, does come with a problem: the "window period." According to the FDA, "up to two months" may elapse between the time of infection and the body's production of the antibodies the ELISA detects. If the ELISA were the only HIV test performed today, I could understand the FDA's erring on the side of extreme caution. But the fact is, three separate HIV tests are now performed on all blood donations—the ELISA, plus HIV antigen and nucleic acid tests, the latter two effectively detecting the virus itself immediately after infection. If done correctly, these tests are accurate.

Louder voices than mine have taken up the cause. Like many people, gay and straight alike, California State Assembly member Mark Leno finds the ongoing ban "blatantly discriminatory," and he has fought to change it for more than four years. Assemblyman Leno told me that back in January 2000, when he was a member of the San Francisco Board of Supervisors, he gathered six men like himself—gay and HIV negative—alerted the media, and headed to the local branch of Blood Centers of the Pacific, the same facility I visited. On camera, standing on the blood bank's steps along with its administrator, Leno called for a change in the policy.

What do you call a protest without a confrontation? Unnewsworthy? Well, no, for this story held a twist: "Even the administrator herself agreed that it was a foolish policy," Leno recalled. "She was frustrated, too. The ban shrinks the available donor pool when instead we need to expand it." In the time since, the problem has only worsened. According to the American Red Cross and America's Blood Centers, which together represent virtually all U.S. blood banks (including Blood Centers of the Pacific), many facilities across the country routinely have less than a day's supply on hand and can't meet hospital demand. While the need for blood steadily increases each year, due in large part to the rise in heart and cancer surgeries, organ transplants, and other complex procedures requiring large transfusions, blood donations are on a steady decline. About 95 percent of qualified blood donors do not give, according to a recent statistic.

To bolster his argument, Leno and his staff did a rough analysis showing that if just one in twelve HIV-negative gay men in the United States donated regularly, their annual contribution would represent one-third of the blood needed every year by the nation's hospitals. Joining forces with the Blood Centers of the Pacific and numerous medical experts, Leno helped lobby for a change in the FDA's policy on gay donors, with the aim of shrinking the over-twenty-year abstinence period down to five years or, better yet, down to one. But the Red Cross fought hard against it. And when it came up for vote in September 2000, the FDA's advisory panel voted seven to six to uphold the ban indefinitely. The years 2001, 2002, and 2003 passed without official debate on the issue. Over time the abstinence requirement, anchored in 1977, grows more punitive.

Shortly after the first vote, I spoke with FDA medical officer Andrew Dayton, a nice guy who carefully defended the agency's position. "We have a strong congressional and public mandate for zero error," he explained. "If we change the policy and something happens, it's a very big issue. We have to be ultraconservative."

Of course, I absolutely understood that great precautions must be taken with our blood supply, but what made sense in 1985 no longer does, given the triple HIV testing done on donated blood. To my mind, the ban perpetuates an early-AIDS-era myth that the blood of gay men is intrinsically different, dirty, or bad, a fallacy that harks back to the ancient belief that the blood contained the essence of a person. I recalled how this misconception had reared its ugly head early in the history of blood banking, during the early 1940s. Posters plastered across major East Coast cities called upon Americans to do their part for the war effort by donating blood—one powerful image showed a wounded GI using his rifle in an attempt to lift himself, with the headline "Your Blood Can Save Him"—except that there was

some invisible fine print: Black blood wasn't always welcome. In Red Cross blood drives carried out in the eleven months leading up to Pearl Harbor, all African Americans were expressly prohibited, as per a new policy established by the U.S. military. As journalist Douglas Starr explains in his book *Blood: An Epic History of Medicine and Commerce* (1998), the armed forces were segregated at the time and "its leaders thought it best for morale not to collect African American blood," the assumption being

World War II blood-drive poster
(Courtesy of the American Red Cross Museum. All rights reserved in all countries.)

that white soldiers would object to having "colored" blood put into their veins. The possibility that some black soldiers might not want Caucasian blood did not figure into this decision. As Starr continues, the policy was "liberalized" soon after December 7, 1941, when the Red Cross successfully lobbied the military to accept blood from black citizens, though it would be processed separately and labeled for use only in "Negroes." Following the war, the institutionalized segregating of blood continued in many American hospitals, particularly in the South, into the late 1960s. Ignored throughout these turbulent times was the perspective of prominent scientists who, one after the other, declared that, in terms of race, blood is blood is blood. The practice was medically baseless.

I launched none of this heavy history at Andrew Dayton during my talk with him because, I must admit, I was hoping that he might have a surprise for me, some stunning revelation to turn my frustration with the gay ban 180 degrees. Well, he did turn it a few degrees. If the FDA policy were changed, Dayton told me, the biggest danger would not be gay donors per se but, instead, the workers handling the blood. The problem would be human beings making human mistakes—the employees who accidentally release HIV-infected blood instead of disposing of it. This already happens, he acknowledged. About ten units of tainted blood products are mistakenly okayed for release in the United States every year, causing two or three HIV infections. "The problem is not the large blood banks," Dayton said, "but smaller blood collection facilities, typically in hospitals, which don't have the staff or automated equipment. They do it manually and have the highest risk of error."

When I asked what the FDA is doing to reduce such errors, Dayton admitted, "It's not quite clear what direction to take." He was unequivocal, however, on one point: "It's important to keep high-risk donors from even giving a unit of blood."

The ban on gay donors conceivably could change, he conceded, if specific research were done. "What we're lacking is seroprevalence rates, the frequency of HIV infections in men who haven't had sex with another man for one year versus five years versus twenty-three." He added, "I think that if we got results that said rates are virtually the same as the general population, then that would put an end to the question." While the FDA has encouraged the Centers for Disease Control and the National Institutes of Health to organize such studies, Dayton noted, none is currently planned, nor is there funding to support them. Even if data were presented and the policy changed, the best-case scenario, he posited, would likely be a five-year deferral for gay men following their last sexual encounter, still far beyond what's required for other groups. In my case, I would never qualify as a blood donor so long as I'm with Steve—and definitely not so long as he has AIDS. And neither of those will change. In the FDA's eyes, Steve's and my realities are the same: My blood's as bad as his.

When he and I first got together, friends were dying of Kaposi's sarcoma, Pneumocystis carinii pneumonia, and toxoplasmosis, all of which can now be prevented or treated. And while protease inhibitors certainly extend lives, they contribute to new health problems—heart disease, lipodystrophy, and kidney or liver dysfunction. Overtaxed organs may finally just give out. Should Steve ever get that sick, I would give my life for him, by which, in practical terms, I mean I'd donate spare parts of my living body—a kidney or half a liver, whatever he needed. And I could—there is no restriction against healthy, HIV-negative, gay men donating organs for transplant. The final irony is, were I to die today, I could literally give Steve my heart, yet when it comes to blood—such a simple, plentiful gift—I am not allowed.

TWELVE

Blood Lust

BLOOD LIVES IN NEAR-TOTAL DARKNESS. WITHIN THE
body it travels along the many thousands of miles of
vessels under the deep shade of bone, flesh, and skin.
Except during its jaunt across the eyes. These red
threads in the whites of the eyes aren't veins but arteries, it dawns on me early one morning. So obvious
once you think about it, the color's the giveaway, the
blood so bright because its cells have just taken a
deep breath. In the same way that the eyes gradually
adjust when you enter a dark room, the closer I study
my reflection, the more blood I begin to see just
under the skin's surface.

The hot water in the bathroom sink has once again fogged the mirror, and I give it another swipe of the hand. In the swath of me, I see the venous blood that purples the circles beneath my eyes, the blue earthworms of my temples. If I shut one tired eye—and oh, how the second wants to follow—I see the web of tiny capillaries on the outside of the lid. It's as though I've showered in luminol, that blood-revealing solution used by crime scene investigators.

Shaving, I try too hard not to cut myself, and I do. Though minor, it's enough to make me flash on a scene that's stuck in my head since my last reread of Bram Stoker's *Dracula:* It's a little past sunrise, a few days into Jonathan Harker's visit to the count's Transylvania castle, and the young man is shaving in his room. He fairly jumps out of his skin as a cold hand settles on his shoulder and Dracula utters, "Good morning," though nowhere in the mirror can the count be seen. Jonathan's nicked himself and the sight of blood running down his chin seems to quicken Dracula's. Only the crucifix hanging at his throat keeps the count from pouncing. "Take care," Dracula purrs before retreating. "Take care how you cut yourself. It is more dangerous than you think in this country."

Thinking about this exquisitely creepy scene makes me realize how differently it would play if it were set in the vampire world created by contemporary novelist Anne Rice. It wouldn't take place during the daytime, for one thing, because Rice's vampires can be injured or destroyed by sunlight. The crucifix, on the other hand, would cause no harm. Nor would the vampire be invisible in the mirror. In fact, since possessing great beauty is a prerequisite of being "turned"—so that the insult to God might be greater, as one vampire explains—Rice's creations might even consider it cruel if denied their reflections for eternity. Also, unlike in *Dracula,* such a scene would never unfold from the mortal's point of view. The reader would be placed inside the vampire's head as he stalks the young man, lusting for his blood while also hating himself for the lust. Finally, while an Anne Rice vampire wouldn't possess the power to slip through a keyhole, which is how Dracula magically snuck into Jonathan's room, one could easily insinuate himself into a prospective victim's bedroom in the more traditional way—through the art of seduction.

As Rice's first vampire book opens, for instance, a mortal enters a vampire's room, rather than the reverse. The young man has been enticed there for something illicit, thrilling: a story. The vampire promises it will be a good one. By all rights, the young man should be terrified. After all, he is alone in a room with an intense stranger he just met in a bar, a predator driven to drink human blood. But instead, the boy is utterly intrigued by this elegant, articulate character, the vampire Louis.

When I moved to San Francisco in 1985, the year the second book in the series came out, the fact that I hadn't yet read the first earned me a joking reprimand from my new roommate, Rich: "Bad, bad homosexual!" as if I were a puppy who'd not been housebroken. He gave me a copy of Rice's *Interview with the Vampire* along with another essential work I'd yet to read, Armistead Maupin's *Tales of the City*, deeming this one of his cultural duties as a gay man who'd lived in the Castro for more than a decade. The two books were night-and-day versions of life in San Francisco. *Tales* was a delightful breeze, set in the 1970s, pre-AIDS, while the lush, dense, and tragic *Interview*, though it had been published in 1976, seemed to have been written expressly for San Francisco of the mid-1980s.

The story of *Interview*, with its brilliantly simple setup, struck a chord with me at age twenty-four. It read like a cautionary tale about dating during an epidemic. In Louis you meet a supernally attractive, urbane man who says he just wants you to know him.

He wants to know you. He invites you back to his place. You go, though you know this guy is dangerous. But he is so irresistible. You spend the night together, locked in a profound intimacy. Oh, the things you talk about. Well, he does most of the talking, but that's okay. You get to stare into those amazing eyes, all the while knowing that if you're not careful, if you let your guard down, he can infect you with what infects him.

I could appreciate Daniel the interviewer's risk-taking for the sake of an extraordinary story. But I also understood Louis's motivation. Though the safety of all vampires lies in each one's silence, for now he doesn't care. A power beyond him has turned him into something he loathes, a monster, and he knows he can never change. He consents to the interview for a deeply human reason, to purge himself of his secrets. For myself, as a young man who had just horrified his parents by telling them I was gay and moving to San Francisco—"You might as well commit suicide" was my father's bon voyage—I saw in *Interview* something instantly familiar. It was a vampire's coming-out story.

Early on in the book, Louis tells Daniel of the anxious final moments of his first night as a vampire. As dawn approached with its killing rays, he'd accompanied Lestat, the vampire who'd "made" him, to a room in New Orleans. Accommodations were spare, so the two men would have to bed together. "I begged Lestat to let me stay in the closet," Louis recalls, but the elder bloodsucker just laughed, exclaiming, "Don't you know what you are?" Lestat slid into the narrow coffin first, then pulled Louis down on top of him and shut the lid. The two would sleep face-to-face. The following evening Louis would awaken and take his final step in crossing over. He'd hunt for the first time and drink the blood of another man.

. . . .

THE DESIRE OF THE UNDEAD TO SUCK AND SWALLOW MOUTHFULS of liquid life has more to do with hunger than with thirst. The blood drive is the sex drive in the world of vampires. In ours, conversely, sex is driven by and dependent upon the blood, which works its own dramatic transformation on us humans. The change begins well before the clothes come off.

Naturally, the impetus for arousal varies from person to person, but regardless of the accelerant—a look, a smell, a touch—the biology is consistent. As ardor takes hold and heartbeats quicken, the brain green-lights the circulatory system to rush blood to certain sexually pleasing places as well as others less obvious. Capillaries in your earlobes and those lining the interior of your nostrils, for instance, will fill with freshly oxygenated blood, causing the skin to plump and become extra sensitive. Likewise, the tiny vessels in the lips and tongue fatten and warm, literally raising the temperature of your kisses.

Though it sure may feel like it, blood doesn't increase in volume during arousal but gets redirected. In women, blood turns the pelvic area into a tropical zone, the labia and clitoris swelling, sensitivity building. The breasts, too, become fuller, the nipples stiffening from the blood-soaked spongy tissue within. Male nipples perform similarly, though, being of smaller stature generally, at a more modest scale. Of course, a grander transformation occurs in the groin, where arteries dilate to allow increased blood flow to the penis. Here, forming the length of the shaft, are three clustered cylinders that dangle like a soggy noodle when the penis is flaccid. (The urethra runs through the bottommost of these.) As these spongy tubes soak up blood, however, the organ bulges in all dimensions—on average, about two extra inches in length, more than half an inch in girth—raising the pressure within until it stands erect.

That it is called an erection merits a wee digression: How

very male and grandiose the word sounds to my ear, bringing to mind such awe-inspiring feats of engineering as hoisting an ancient obelisk or raising a modern skyscraper. In point of fact, achieving an erection requires less blood to the penis than one might imagine, though don't tell this to your typical size-sensitive male. About two ounces—or, one-eightieth of a 150-pound man's total blood volume—is all it takes to make him hard.

Leonardo da Vinci

From the classical age to the Renaissance, it was believed that an erection was due to a breath-like substance brewed in the liver, Natural Spirits, which inflated the penis as, to use a modern analogy, air does a tire. The brilliant Leonardo da Vinci, a visionary in conceiving of such marvels as flying machines and diving gear, was also prescient in identifying the inner workings of male genitalia. More than a hundred years before blood's role in erection was first correctly described in Western medical literature, Leonardo accurately summed it up in one of his illustrated notebooks. In 1477 he'd attended the public hanging of a criminal in Florence and, like others in the crowd, couldn't help noticing that an erection was a consequence of this form of execution. During the subsequent dissection of the man's body, Leonardo saw that it was in fact blood that had filled the organ, a result of the violent, downward jolt. (Incidentally, the phrase *well hung,* slang for "having large genitals," does not derive from such observations. Rather, it dates back to an early-seventeenth-century description of a man's jumbo ears, of all things, a usage that soon broadened to encompass any oversized body part. In any event, grammatically speaking, a noose and a fall lead to being well hanged, not well hung.) Following the dissection, Leonardo wrote, "If an adversary says wind caused this enlargement and hardness, as in a ball with which one plays, I say such wind gives neither weight nor density. Besides," he added, referring now to the head of the phallus, "one sees that an erect penis has a red glans, which is the sign of the inflow of blood; and when it is not erect, this glans has a whitish surface."

In his work *A Mind of Its Own* (2001), a "cultural history of the penis," David M. Friedman writes with an implied wink that Leonardo, whom modern scholars agree was homosexual, "investigated the male member as no one before him ever had." He filled page after page with detailed anatomical drawings along with quirky observations. Leonardo noted, for example, "The woman likes the penis as large as possible, while man desires the opposite of the woman's womb. Neither gets his wish." (I'll assume Leonardo came to this conclusion anecdotally.) Further, the male reproductive organ, to his eye, was ideally situated on the firm base of the pubic bone. "If this bone did not exist," Leonardo hypothesized, the penis during intercourse "would turn backwards and would often enter more into the body of the operator than into that of the operated." In other words, one would end up screwing oneself.

Once blood was correctly implicated in erections, new discoveries were made and new misconceptions arose. Dutch scientist Reinier de Graaf, history's next great investigator of the penis, correctly documented in 1668 that the penis does not, in fact, contain a single ounce of fat. What you see is lean flesh and blood. Its size will not change with weight gain or loss. De Graaf was also correct in declaring that the key to maintaining an erection is not getting blood into the penis but *keeping* it there. Alas, his theory that trapping blood in the penis depended on muscular constriction was interesting but wrong, as was the contention of subsequent scientists that valves in the blood vessels did the job.

Not until the early 1980s did the actual mechanism come to light. As it turns out, the process depends on what sounds like a physiological contradiction: A man gets hard because a crucial part of the penis softens. With the sudden influx of bright red arterial blood, the smooth-muscle tissue lining the three cylinders of the penis relaxes and, as a result, expands so fast that the veins through which blood normally returns to the heart get flattened against the shaft's outer walls. In effect, a hematological flash flood has taken place, leaving all exits blocked. (A separate mechanism impedes urination, freeing the urethra to transport only semen.) With circulation cut off and the store of oxygen waning, the penis darkens in color, just like when you tighten a rubber band around a finger. The floodwaters typically will retreat soon after ejaculation. However, in the condition called priapism—named for the Greek god of virility and sexual prowess, Priapus, whose endowment, shall we say, was legendary—erection persists well beyond the point of enjoyment. Brought on by certain medications, injury, blood disorders such as sickle-cell anemia, or, in many cases, by reasons leaving doctors scratching their heads, priapism is painful and becomes dangerous if it lasts more than four hours. If the penis is not decompressed, the trapped blood starts to clot and must be extracted using a remedy that will make any man wince: A large needle is inserted into the shaft and the thickened, almost black blood is sucked out.

To the other physiological extreme is erectile dysfunction, for which several well-advertised treatments are available. Drugs such as Viagra and Cialis, contrary to popular belief, neither increase libido nor trigger an immediate erection. Rather, they depend on a key ingredient not in the pills: arousal. But once that's fired up, Viagra, for example, stimulates the release of a chemical that increases blood flow to the penis while also inhibiting an erection-wilting enzyme.

Now, returning to women: The clitoris, unlike the multipurpose penis, exists solely for pleasure. I can think of no other body part where so little blood does so much good. Engorgement exposes the clitoris, normally hidden within the vulval cleft, and amplifies the sensitivity of its eight thousand nerve fibers—twice the number found in the entire penis, in a much smaller area. Though blood is the common agent, erection and engorgement are significantly different processes, as science writer and author Natalie Angier clarifies in *Woman: An Intimate Geography* (1999). Because the clitoris does not share the penis's distinctive outer plexus of veins, its vascularization is more diffuse. Hence, when the clitoris swells—typically to twice its size—the outflow veins are never compressed, and so "the organ does not become a rigid little pole," as Angier cheekily points out. This free flow of blood may be what allows the clitoris to relax and distend again and again, she adds, giving rise to the multiple orgasm.

You must be *brought to* it. It must be *reached.* And in order to be fully felt and appreciated, an orgasm depends on the hot, twisting passageways of the bloodstream. In the build to climax, the hormone oxytocin is launched into the blood in a double volley from the brain and either ovaries or testes. The surge peaks at orgasm, reaching up to five times the normal concentration. By causing heart rate and blood pressure in the typical person to double, oxytocin accelerates its own speedy travels through the body. Most pleasingly, it helps trigger the pelvic shudders that women experience during orgasm and possibly the muscular contractions in men.

Love poets have long rhapsodized how souls touch during lovemaking, and oxytocin may well be the biochemical basis for this claim. The hormone, known to be vital in forging the adamantine bond between a mother and her child (and also that between father and child, probably), may perform a similar function in sexual partners, researchers believe. Oxytocin-spiked blood stirs immediate feelings of connectedness to the person with whom you are intimate, which may either lay the foundation for long-lasting ties or, if you're already well acquainted, strengthen existing ones. In a remarkable sense, then, oxytocin is part of the blood's formula for building familial bonds with those who aren't blood kin.

But blood at orgasm is not just relationship-minded, so to speak. At the same time it helps curl your toes, oxytocin signals other chemicals to gush forth, such as potent opiates aimed more at dulling sensation than amping it. These are of the same family as the endorphins released during exercise and have, for example, the effect of temporarily numbing raw nerve endings in people with migraine, arthritis, or peripheral neuropathy. Furthermore, oxytocin activates mechanisms that help heal wounds and raises blood levels of immunoglobulin, a microbe-fighting antibody. So significant are these and other benefits that, experts agree, not only is sex good for you, but it may also lead to your living longer.

Of course, those same experts are not saying, when you're sick, have lots of sex. The heat of passion and the heat of illness do not usually overlap. And for good reason. When you're sick, the blood is rigged to make you drowsy. Hypnos, the god of sleep, almost always thwarts Eros. You need only have a cold to appreciate this. Or, for the sake of illustration, say you've got a flu bug—nothing too serious, but enough to confine you to bed. While you rest, white blood cells directly attack the infection and, launching a broader assault, the body changes its internal environment to make itself less accommodating to invaders. Cellular messengers called pyrogens—"fire starters," roughly translated—are sent through the blood to the body's thermostat, the anterior hypothalamus in the brain, which turns up the heat. At the same time, blood vessels in the skin narrow, reducing sweating, which is a major way body heat escapes. Now producing more heat than it can lose, the body runs a fever. (By the way, for most people, what's widely considered the normal body temperature, 98.6 degrees Fahrenheit, actually isn't. An enduring math error from the 1800s is the cause of this misconception. The true average is 98.2.) Fever, which helps kill whatever virus or bacterium swims within, may be "a lovely way to burn," as Peggy Lee crooned, but, truth be told, it's lousy for lovemaking.

Even if there weren't such sound immunological explanations for the lost carnal itch, common sense should tell you plenty: You could be contagious. You shouldn't overexert yourself. You look like the unfed undead. And naturally, your sex appeal suffers, too. But once you or your bedmate have bounced back to health, desire follows. I'm sure almost anyone can appreciate sex after illness—the first romp, say, after your once-yearly flu—but I believe you have to have known serious illness or injury to truly savor it. As a broken bone is said to be stronger once mended, so, too, is lust restored.

Many times in Steve's and my relationship, our sex life has had to be packed away. His meds have often been to blame. One drug dried his skin so terribly that his lips bled, making kissing out of the question, and others, designed to free him from physical pain, distanced him from good sensations. Over the years many of his prescriptions have come with the warning label THIS DRUG MAY IMPAIR YOUR ABILITY TO OPERATE MACHINERY, without ever noting that the machinery included his own.

But the body usually gets used to drug side effects over time—if there's time. The most difficult period for Steve came during our third and fourth years together. During a long bout of wasting syndrome, he'd steadily lost a frightening amount of weight and, as he now reflects good-naturedly, his libido went down the toilet, too. On his doctor's orders, I began giving him regular injections of testosterone enanthate, not to help him reclaim virility but just so he could hold on to some of his mass. Still, he got so thin that he could no longer wear the ring I'd given him on our first anniversary. Everything began to change for the better, finally, thanks to new medications. Steve's T cells rose, he gained weight, the color came back to his face, his appetites returned. We had no idea how long he'd remain stable, which, as I see now, gave sex an intensity that was bittersweet. *This might just be a reprieve,* I thought, just a short break of sun. To this day I've never been able to shake that feeling.

The first time back out can be a little awkward at first. Naked, you feel unusually exposed. Skin is the body's largest organ, a marvelous complex of nerve endings, sweat glands, and the tiniest of blood vessels, the capillaries. At any given time, about one-quarter of the body's blood flows through the skin. Even so, it may take a moment to warm up, to get the blood moving. My partner and I remove our clothes and reach for each other as if underwater, two bodies meeting at the bottom of a pool. Pushing against resistance, we kick our legs to stay in place, hold our breath, close our eyes. Just as we make contact, we surface, mouth to mouth among the waves.

Memory Cells

SHORTLY AFTER STEVE AND I BEGAN LIVING TOGETHER IN early 1990, we cemented our coupledom by making deposits into a joint savings account. The money wasn't earmarked for a future vacation or a down payment on a home. This was our cure fund, a stash of cash set aside for the day the magic bullet would be discovered. We knew Steve's health insurance wouldn't immediately cover a brand-new treatment, no matter how miraculous. And for some strange reason, we both felt sure the cure would be found overseas. On a moment's notice, we'd have to board a plane to who knows where. Steve thought it might be Japan. I thought France. Hungry for news, we attended the monthly Project Inform updates held in a church in the Castro. As in an old-fashioned town-hall meeting, anyone could stand up and speak, share treatment success or horror stories, or ask questions of the evening's guest—a visiting doctor, usually. Oftentimes, though, lively debate devolved into medico-babble few of us in the creaking pews could follow.

I remember how, in June of that year, everyone was talking about an "amazing" and "promising" experimental therapy called hyperthermia, or blood boiling. Sure, it had only been tried on two people, but the basic science behind it sounded pretty solid. The opposite of hypothermia—the condition suffered by people who plunge through thin ice—hyperthermia mimicked a high fever, the body's infection-fighting mechanism. The procedure involved withdrawing the blood from an AIDS patient's body, pint by pint, using an apparatus that was a crude forerunner of today's blood bank e-chair. The blood was heated to up to 115 degrees Fahrenheit, thus killing the virus, then cooled and returned to circulation. The procedure, performed by an Atlanta doctor, took a mere two hours. One of his two treated patients went on national TV and declared himself cured of AIDS—an overstatement, yes, but his viral activity had in fact dropped significantly and his T cells had shot up. Soon thereafter, though, the blood boiling fever broke. A third patient died from the procedure, and government investigators promptly deemed hyperthermia dangerous, worthless, and done with. As it turned out, what we'd all conveniently forgotten in the excitement was that HIV takes up residence not just in the bloodstream but in organs and the glands of the lymph system. So the blood would inevitably be flooded with new virus; it was just a matter of time.

In the spring Steve had started taking AZT, the sole FDA-approved drug at the time, and he'd begun seeing a new physician, a Spanish-born woman recommended by a friend at Project Inform. Our first impression of Dr. Inmaculada Marti was that she had turned what would've been just another drab office in the Davies Medical Center into a glistening cavern. The space was a geode—the shelves, windowsill, and her whole desk covered with crystals, save for a tiny spot reserved for her prescription pad. Steve stuck with her for about nine months. She recommended acupuncture, which he tried, and at every office visit, while I watched from the rose quartz section, she would spend a lot of time examining his tongue. As his immune system staggered, Dr. Marti seemed to get angry first at the lab results, then at Steve, as if he were an uncooperative patient. "You're taking an antiviral," she bristled during one appointment, "so why do you have virus?" She started him on an alternative therapy, Iscador, an extract of mistletoe, which she ordered from Switzerland. In the end, however, we knew she'd run out of tricks when she proposed sending a sample of Steve's blood to New Mexico, where, for four hundred dollars in cash only, a colleague would perform a "visual study" of his virus and cells. How would the equivalent of a tarot card reading of Steve's T cells help steer his treatment? Dr. Marti admitted she couldn't say.

After switching to a new doctor, a wonderful woman firmly grounded in Western medicine, Steve, over months, then years, worked his way through the latest antivirals—ddC, ddI, d4T, 3TC—all bleached-white tablets, like generic aspirin. By contrast, I remember how very different Steve's first protease inhibitor looked. Saquinavir came wrapped in bright gold-and-green capsules, plump and shiny like movie candy.

I once read an article on how pharmaceutical companies create the names for their forthcoming products, a skill requiring sales savvy and a gift for poetry. Experts invent words aimed to evoke just the right feeling, mood, quality, or image. The best-named drugs seem to start working just by saying their names. Say it slowly again and again and the sleep aid Ambien becomes a lulling chant. And it's no coincidence, I'd wager, that Viagra is so close to Niagara, that mighty flow of fluid. To me, the names given to the protease inhibitors seemed intended to call to mind fabled heroes of an earlier time. One after the other they arrived, modern-day Knights of the Round Table: Saquinavir, Ritonavir, Nelfinavir, Crixivan.

In the Arthurian legend, as Steve recently pointed out to me, more than a hundred knights earned a place at the Round Table, but a single seat, or "siege," remained empty: the Siege Perilous. It could be filled only by the true finder of the Holy Grail and would bring death to any pretender. Though the notion of killer furniture seems silly to me, I do love the element of the waiting chair. This says so simply that the right person just hasn't arrived yet. One day, the Siege Perilous will be occupied.

I SPOKE WITH DR. JAY LEVY ALMOST TWENTY YEARS TO THE DAY after he had co-discovered the virus that causes AIDS. When I first mentioned the milestone, though, he seemed a little surprised at the math. "Oh, that's right," he said, brightening, doing his own mental tally. "That's right, twenty years." Levy, who heads the Laboratory of Tumor and AIDS Virus Research at the University of California, San Francisco, added that he'd actually made the September 1983 discovery in a lab right across the hall from the small office where we were seated. But, he noted, it's now gone.

I leaned back in my chair and glanced across the hallway, compelled by that queer human reflex to look for something we're told is not there. *What is it that we think we'll see?*

"They took the room," Levy added.

"Someone 'took' the room?"

"The Smithsonian, yeah." A bemused twinkle lit his gray-green eyes. "They came in and just took the whole thing, where we'd isolated the virus. The lab hood, my lab jacket, my note-books, everything. The sign on the door. One day it's going to be in the Smithsonian."

As he continued the tale, I realized that, to him, the furniture and equipment hadn't been so much historical as old and in the way. What seemed to truly please the sixty-four-year-old scien-tist was that the university renovated the lab, providing him a tidy new space for his ongoing research.

Finding HIV was one thing. "Conquering this virus," as he put it, has occupied him since. In the mid-1980s, for example, Levy developed a technique for inactivating HIV in the clotting factor preparations used by hemophiliacs. "Then I became sort of an expert with blood products—*How do you get rid of HIV without destroying the proteins you're dealing with?*" It's a good indication of the man that he modestly describes this work as "helping out." Levy's pioneering heat treatment method, adopted by the blood products industry, has saved many lives.

Levy, author of the seminal text *HIV and the Pathogenesis of AIDS* (1994), among other books, was also the first scientist to report that HIV could cross the blood-brain barrier, the filtra-tion system that normally protects the brain from harmful sub-stances carried in the bloodstream. To me, the term *blood-brain barrier* had always given the impression of a single dam-like structure at the base of the skull, where I'd imagined it was lo-cated, but the barrier is actually the layer of tightly packed cells that make up the walls of all brain capillaries. Levy correctly concluded that HIV's ability to pass through this barrier led to neurological diseases such as AIDS-related dementia. Since some of the earliest available anti-HIV drugs did not cross the blood-brain barrier, his findings underscored the necessity of developing drugs that could.

I had met Dr. Levy in person once before today. In this same building, eleven years earlier and twelve floors below, standing before six hundred people in a five-hundred-seat lecture hall, he and two other prominent researchers had spoken at a public forum I'd organized as a staff member of the San Francisco AIDS Foundation. The free event was the kick-off for my 1992 "Be Here for the Cure" media campaign. Like the campaign, the forum was a look forward. The topic: prospects for a cure. As the event was about to get under way, I adjusted the lectern microphone and scanned the packed auditorium, a great throng of gay men, mostly. People stood in the back, jammed the aisles, and sat on the floor before the stage where Marcus Conant, the pioneering AIDS physician; Mathilde Krim, cofounder of amfAR; and Levy were seated. I spotted Steve, who was saving a place for me on the stairway, and went to join him.

Both the setting and the stature of the speakers gave this gathering a different feel from the frantic town-hall meetings of two years earlier. The mood had changed, too. It was as if, finally, science had given hope a scaffolding to build on. Dr. Conant spoke about a new treatment strategy, "combination therapy," now standard protocol, and reported how taking prophylaxis against Pneumocystis pneumonia increased life expectancy. Dr. Krim talked about promising vaccines under development. But the night's big bombshell came from Dr. Levy. "I do not believe that everyone infected with HIV will come down with symptoms," he told the crowd. He then described a "blood factor" he was investigating and predicted the day when a devastated immune system could be restored to full fighting strength.

Dr. Levy remembers that evening well, though, as he admitted to me in his office, he has never really liked the word *cure*. "*Cure* is not the word. *Control* is the word," he stressed. Though significantly grayer than when I'd last seen him, Levy spoke with the same certainty. "You aren't gonna kill every damn cell affected in the body. HIV is a virus like herpes that you have forever. But we *will* regain control of it."

To reach this end, Dr. Levy has spent the bulk of his energy over the past seventeen years studying a small pool of research subjects: those rare individuals with HIV known technically as long-term nonprogressors or, more commonly, long-term survivors. Statistics indicate that only about 1 percent of the infected fall into this category.

"We first started to look at these people back in 1986 and found that the virus in them couldn't replicate," he recalled. The subjects also had relatively normal white blood cell counts. In those early days, without drug therapy, most people diagnosed with AIDS declined rapidly. "But these were healthy people who showed no signs of infection. And we asked, *What's the secret?*"

For answers, Levy, a trained virologist, crossed over to a new turf, immunology. "One of the things with viruses," he explained, "is that, to survive, they have to exist within a hostile environment." The brunt of this hostility comes from the body's natural antiviral response. After HIV invades, antibodies produced by B cells, the fleet of T cells, and other weapons of immunity flood the bloodstream, mounting a formidable defense. But inevitably, in nearly every case, the immune system is soon turned against itself as viral RNA commandeers helper T cell DNA, or, simply put, the helper cells become factories, churning out more copies of HIV.

Not, however, in the bodies of nonprogressors.

In these lucky individuals, Levy found, the internal environment remains antagonistic to HIV, not simply for weeks or years but, in some, for decades. On this topic, I realized I had been operating under a misconception. By virtue of these survivors staying symptom-free, I'd assumed that, within them, all was weirdly peaceful—a milieu wherein HIV and white blood cells quietly coexisted. *Don't bother me, I won't bother you.* Not true. Nonprogressors' immune systems actually contain supersoldiers whose secret weapon, Levy discovered, is a chemical substance he named CD8 Antiviral Factor, or CAF.

CAF is a protein produced by a type of T cell called the CD8 lymphocyte, or suppressor cell. (These white blood cells don't "suppress" invading organisms but, rather, inhibit the activity of fellow cells within the immune system.) As Dr. Levy explained, while CAF does not stop HIV from infecting helper T cells, "It blocks the virus so it can't make its RNA. HIV is shut off by this factor."

Everyone has CAF, I was glad to hear him say; however, he believes that "only long-term survivors have a capability of maintaining it." The hope for HIV progressors—my term—is learning how to provoke or rekindle CAF production and maintain it over time.

But first, Jay Levy must actually find this elusive protein. While prominent AIDS researchers agree with him that an antiviral factor is produced by CD8 cells, Levy has yet to isolate it. On the one hand, he and his team can extract a sample of blood from a nonprogressor that contains this factor and can demonstrate in the lab that it does in fact suppress the virus. But where exactly is it? "I can't yet give you the molecular structure of CAF," Levy said evenly, the very picture of unruffledness. It is this quality, I realized at that precise moment, that separates the true scientists from the test-tube throwers. Since 1989 the Levy team has been on a molecular scavenger hunt, winnowing the possible proteins from a group of two thousand candidates down to two hundred and now down to fourteen. "These are fourteen of our best. Which, potentially, could be CAF. I don't know. But when we find it, it's going to be such a powerful antiviral drug and such an unusual and universal means of bringing back control to the infected person."

I gave him a look that asked, *Should we be holding our breath?* His reply was at once apologetic and resigned: "The fact is, if we were a [biotechnology] company, we'd have the research all done by now."

At this point in our conversation, we left his office and stepped into the adjacent lab, a functional space, but, as he accurately described it, "far from plush." He noted that, though he'd prefer to spend all his time in here working on research, every year the struggle to obtain government and foundation grants gets tougher. "I spent 50 percent of my time last year traveling around trying to get the funds." He sighed his frustration. "It's not right."

The basic science required to hunt down CAF offers no guarantee of a quick return on investment, he continued; it's just too "high risk" for a biotech company to consider. Enter irony: Levy has received offers to become the head of a few such companies, to which his reaction has always been, "But then who's going to work on CAF?"

Levy agreed to show me his other work spaces and led me down hallways that interlocked with the puzzling aspect of a hedge maze. Finally ushering me through a door, he said with good humor, "You need roller skates to get from one lab to another," though, I must say, Levy is not a man one can easily imagine on skates. A Segway, however, is a distinct possibility. He motioned to a glass-front refrigerator, and I peered in at a fleet of vials. "In here, we've got cells that are growing the virus. In the old days," he added dryly, "no one would come in this room."

"In the old days, no one would come in this room." *Jay Levy in his laboratory, 1984*

In a connected lab, Dr. Levy pointed out two members of his staff, impossibly young-looking postdoctoral fellows, Leyla and Hillary. They gave a quick, friendly wave. "Leyla's the one who narrowed it down to fourteen proteins," he said with pride. "It took her four years." She smiled modestly and I thought, *Remember that face: You might just see it on the cover of* Time *someday.*

Although he and his staff hope to get through all fourteen in the next three years, Levy conceded, "We may end up finding we missed it." His entire body language answered my unasked question: Yes, the hunt would then start over.

As we were stepping out the door, a splash of color caught my eye, a familiar poster, BE HERE FOR THE CURE. We'd given away thousands of these back in '92. I smiled and gestured to it: "That looks good." Levy said it had been up since the forum. The poster, I noticed, hung right where they turn the lights on every day—*click*—and where they turn them off.

. . . .

A FEW DAYS LATER I VISITED CHIRON CORPORATION AND SPOKE WITH a former senior member of Levy's team, Dr. Susan Barnett. Chiron, whose U.S. headquarters is in Emeryville, just across the Bay from San Francisco, is a global biotech company with a legendary namesake: Chiron was the centaur who'd taught the healing arts to Asclepius, the Greek god of medicine and father of the goddess of cure-alls, Panacea.

Barnett, an effervescent forty-eight-year-old, is Chiron's project leader for HIV vaccines. She greeted me in the soaring atrium of the life sciences building and led me up a broad stairway to her sleek, uncluttered office. I happened to be meeting Barnett on a day when she was experiencing a heady milestone of her own. In a phone call, she'd just gotten "the thumbs-up" from the FDA for the first Phase I clinical trials of their new HIV vaccine, a product Chiron's been developing for almost ten years. I do believe that, had Barnett had a confetti cannon, she'd have set it off right then and there. "I'm as excited as a scientist can get," she admitted. "We're on the precipice," she added with a warm smile, although her imagery clearly evoked the perils inherent in taking the leap into human testing. Will it fly or fail?

The company has been in a similar precarious spot before, as Barnett recollected. Back in 1994, when she'd arrived at Chiron from Jay Levy's lab, the first generation of AIDS vaccines was about to enter Phase II trials, in which efficacy is tested. Under study was not only Chiron's vaccine but also vaccines developed by other pharmaceuticals, all employing a similar design: the use of a single small part of the "envelope" of the virus.

"And they didn't go," Barnett recalled. Indeed, based on dismal Phase I results, the government pulled the plug on this approach to vaccine creation, deeming it hopelessly ineffective.

"That was a pivotal moment," she told me. "Everybody retrenched and started doing more research." Her voice had grown sober in the memory. What was a major setback for companies such as Chiron was also a devastating blow to communities and individuals affected by the epidemic. So, yeah, I agreed with Susan, 1994 was not a banner year.

Chiron's latest, best hope is as different from the failed version as a hybrid car from a 1970s gas guzzler. "We've combined everything we have, our best weapons: DNA technology plus protein technology," Barnett explained, her enthusiasm renewed. The vaccine, designed to protect uninfected people, is administered in two parts. First a series of "prime" immunizations, which is the DNA aspect of the vaccine, at zero, four, and eight weeks, followed by the protein "boost," probably at twenty-four and thirty-six weeks. A promising concoction is only the beginning of a good vaccine formula, I was coming to realize. The true finesse comes with the administering and the dosing. As Barnett described the protocol, I couldn't help but hear an echo from Jay Levy. "Vaccinology is an art," he'd said to me. And now, standing before me, dry-erase marker in hand, was an artist.

At this point Barnett dived into a finer presentation of the relevant mechanisms, complete with visuals, all the while checking in with me with a look: *Okay? Okay? Still with me?* Although phrases such as *trimeric proteins, polymer microparticles,* and *occluded loop* whizzed past me like bullets in a Yosemite Sam cartoon, I had no trouble grasping the principle upon which the vaccine's potential success rests, an immutable, biological truth: Blood remembers illness.

Existing within our bloodstreams is an elite class of white blood cell—they can be either B or T cells—whose prime function is to retain a "memory" of infectious organisms the body has previously encountered. Should one of these organisms return, memory cells, as they are called, recognize it and rally fellow B and T cells to fight off the invader. This is why we do not get chicken pox a second time.

Memory cells are a vaccinologist's ally because they can be tricked. A vaccine, in other words, can plant false memories. If Chiron's AIDS vaccine is a success, it will convince memory cells that they've already encountered HIV when they have not. (Chiron's product contains a noninfectious facsimile of portions of the virus.) If a vaccinated person is then exposed to actual HIV, the memory cells will, in theory, initiate a massive immunological response, killing the virus. This expectation is fairly well grounded. The immune system's "secondary response" is usually faster, larger, and more specifically tailored than the first, or primary, immune response. And, again, in theory, the protection should last. Unlike suppressors, killers, and other white blood cells, memory cells live for decades before dying.

Those are some pretty big ifs, I had to admit to myself, especially since HIV can mutate so rapidly and hides so well in the body. It takes only one—just one—infected T cell to restart a cascade of viral replication. "So," I asked Susan, "what does your gut say?"

"I think we're going to get good immune responses. That's my gut. And yet," she added, "I'm not stopping research until I see some human data." Even in the best-case scenario, she noted, it would be six to ten years before licensure.

I remembered how, ten or twelve years ago, researchers often bridged the notion of a preventive vaccine with a "therapeutic" one, meaning that what would protect the uninfected should also aid the infected. Did this still hold water today? Mentioning my partner, I asked Barnett, "Could this vaccine be used in people with HIV as well?"

She put down the marker and returned to her desk, quiet every step. She sat, then said, "It's complicated."

It's unlikely, in other words. And, as she went on to clarify, it's simply not part of Chiron's plan. Of course, there is a scenario in which the vaccine might be used in a person with HIV: if a trial participant becomes infected. As I knew, slip-ups will be necessary for determining ultimately whether an AIDS vaccine is a success, although this is a harsh reality that Barnett and I did not discuss.

Barnett had worked with Jay Levy for four years and counts him as a valued colleague but she is also an outsider, so she seemed the right person to ask, "What's your take on CAF?"

"Well," she said, stretching the word almost to its breaking point. "What Jay's been going toward, it's a noble quest." She leaned forward in her chair. "His work on CAF has been seminal. He showed that this factor specifically inhibited HIV." Furthermore, other scientists, following Levy's lead, began searching for CAF and discovered something unexpected, chemokines, chemical substances that also inhibit HIV. Even more important, Barnett emphasized, they discovered why chemokines inhibit the virus: They bind to the same receptors on the T cell as HIV does, so, in essence, it cannot "dock."

"Now we know *where* HIV binds," she said, her voice and eyes sharing the same fire. "Now we know *how* HIV gets into the cell." She let that steep for a moment, then recapped: "So we went from his initial discovery to discovering beta chemokines to discovering the receptors for them to discovering how the virus gets into the cell.

"And for me, designing a vaccine," she added, "if I know how the virus binds, then I can figure out how to block it." She sat back, smiling, gently shaking her head in wonder. "So that is pretty amazing. He doesn't have the factor, but look at all the research it has stimulated."

I liked her use of *noble quest*, but I doubted Jay Levy would've liked its ring. Built into the phrase is an acceptance that the destination may never be reached, that the journey may be all. When I'd spoken with the man three days back, doubt never once entered the discussion. "When we find the factor," Levy had said, "it's gonna be a hell of a discovery." *Amen,* I'd thought. "People will say, 'God, they spent twenty years on this thing.' Well, that's not unusual—factor VIII took years to be discovered. And penicillin! Interferon's another example. That's what's going to happen here."

He added, "That's what keeps me going."

"Keeps you on the odyssey?" I said.

"That's right," Dr. Levy replied. "I haven't met the Muses yet."

FOR ME, IT'S NOT THE MUSE'S CLEVER WEAVING OF THE TIME-TOSSED tale or the hero's cunning escape from the Cyclopes or even his sweet reunion with his wife after twenty years away. No, what brings me back to *The Odyssey* time and again is a passage right at the halfway point in the story, when Odysseus and his surviving crew are so utterly, epically lost that their only hope of ever finding home is to journey into hell itself to ask directions of a dead, blind prophet. Directions from a blind man. That doesn't sound promising.

Considering all they've gone through since leaving Troy, their descent to the underworld isn't particularly perilous. There Odysseus stands at a rocky pinnacle where the River of Flaming Fire and the River of Lamentation meet, and he carves a narrow trough in the ground at his feet. Following careful instructions, he performs a lengthy ritual that culminates in his pouring fresh blood into the pit. Souls instantly swarm, craving a drink, a ghostly multitude of old men, unmarried youths, "once-happy girls with grief still fresh in their hearts," and a "great throng of warriors killed in battle," including one of his own fallen men. Their faces are contorted, desperate. Though Odysseus finds the sight heart wrenching, he brandishes his sword to keep them back. The prophet Teiresias must drink first of the blood.

Once sated, Teiresias does indeed prophesy a safe route back to Ithaca and, looking far ahead in the hero's journey, adds, "Death will come to you far away from the sea, a gentle Death. When he takes you, you will die peacefully of old age." With that, the prophet withdraws and Odysseus is free to leave this wretched place, to return to his ship and head for home, where his beloved awaits. And yet, in what I've always seen as an act of great kindness, he lingers and allows a number of the other souls to taste the remaining blood. Whereas the vital fluid had restored the prophet's ability to see the future, for these souls it has the opposite effect: Blood restores memory. One by one, they come, drink of it hungrily, and share their recollections of life on earth.

Odysseus and his men flee the Cyclopes

In time Odysseus does make it back to the land of the living, although his return is not met with cheers. "What audacity," the goddess Circe barks at him, "to descend alive into the house of

Hades! Other men die once; you will now die twice." She makes this sound like a curse, which has always struck me as odd. But then, I remind myself, what flows through Circe's veins—ichor, the blood of the gods—gives her immortality. To her, death is but an abstraction. To a mortal, however, twice dead would mean twice alive.

A series of conversations Steve and I had not long ago with a good friend on the East Coast centered on a similar theme. At age seventy-one, Maurice confided one afternoon that he didn't expect to reach his next birthday. True, he wasn't in robust health—he'd had a major heart attack some years earlier and had recently been hospitalized with a dangerous blood clot—but his prediction had less to do with a particular diagnosis than a gut feeling. His father and his older brother had never reached seventy-two, both having died just shy of that age. Why should he be spared? As the day approached, his weakening health seemed to portend that he was indeed on schedule to expire. But the Fates decided otherwise, choosing not to snip his life's thread. Tuesday followed Monday, and, wonder of wonders, Wednesday showed up, too. Maurice woke up and found himself in blue-skied Elysium, somehow relocated to northern Connecticut.

"I'm living my afterlife," he told us a few days later, sounding healthier and more joyful than he had in years, "and I'm going to treat it with great respect." Steve was on the exact same page. He, too, had never expected to reach a milestone, forty, yet he celebrated his fortieth birthday in relatively good health last April. I just quietly listened in on these two survivors—one young for his years, the other old beyond his—as they laughed at their windfall. Then a wonderful realization came to me: As a couple, Steve and I have only just started living our afterlife. We may not even be halfway through our story.

. . . .

INSOMNIAC THAT I AM, WHEN DEATH DOES COME TO TAKE ME, chances are I will not be asleep. I'm okay with that. I'd prefer to be alert to the last, conscious of every footfall.

If I should die of natural causes, I will actually undergo two deaths, I've come to understand. As clinical death nears, your breathing becomes shallower, you fade into unconsciousness, and, finally, your heart, blood, and respiration come to a stop. Is your last clear thought one of terror as your inner motion slows? Or is it of ultimate peace, the sense that, in stillness, the soul can burst free?

Despite the seeming unambiguousness of the term, clinical death is a reversible state, a transition between life and death. But if circulation and breathing are not quickly restored, you progress to brain death, from which no return is possible. By this point, blood has already begun to settle, prey to gravity.

With both hands, I make tight fists, impeding the blood flow into them. My knuckles go white, the skin actually cools, then, after a minute's squeeze, I release. The blood rushes in, a red tide pinkening my palms. The seams darken, the twin ridges of yellowy callus blush. Last, I feel tingling, circulation returning to my fingers. The blood echoes my heartbeat, each finger drumming from within.

References

General

The following sources are used repeatedly through the book. References for individual chapters are listed below.

Blood: Art, Power, Politics, and Pathology. Edited by James Bradburne. Munich, London, and New York: Prestel Verlag, 2001.

Bulfinch, Thomas. *Bulfinch's Mythology.* New York: Modern Library, 1998.

Encarta Encyclopedia. Standard Edition. Microsoft: 2002.

Friedman, Meyer, and Gerald W. Friedland. *Medicine's 10 Greatest Discoveries.* New Haven, Conn.: Yale University Press, 1998.

Graves, Robert. *The Greek Myths: Volumes 1 and 2.* Revised Edition. New York: Penguin Books, 1960.

Miller, Jonathan. *The Body in Question.* New York: Random House, 1978.

Nuland, Sherwin B. *The Wisdom of the Body.* New York: Alfred A. Knopf, 1997.

Starr, Douglas. *Blood: An Epic History of Medicine and Commerce.* New York: Quill/HarperCollins Publishers, 2000.

Wintrobe, Maxwell M. *Blood, Pure and Eloquent.* New York: McGraw-Hill Book Co., 1980.

———. *Hematology: The Blossoming of a Science.* Philadelphia: Lea & Febiger, 1985.

Wintrobe's Clinical Hematology. Tenth Edition. Edited by G. Richard Lee, M. D., et al. Baltimore: Williams & Wilkins, 1999.

One

Aristotle. *Parts of Animals (De Partibus Animalium).* Translated by A. L. Peck. Cambridge, Mass.: Harvard University Press, 1955.

Encyclopedia Mythica Web site. September 2001 and May 2002. www.pantheon.org.

Gods and Heroes of the Greeks: The Library of Apollodorus. Translated by Michael Simpson. Amherst: University of Massachusetts Press, 1976.

Hornik, Susan. "For Some, Pain Is Orange." *Smithsonian* (February 2001): 48–56.

TWO

Brain, Peter. *Galen on Bloodletting.* Cambridge: Cambridge University Press, 1986.

Cozzo, Rosemary. Interview with author. San Francisco, Calif., April 19, 2002.

Davis, Audrey, and Toby Appel. *Bloodletting Instruments in the National Museum of History and Technology.* Washington, D.C.: Smithsonian Institution Press, 1979.

Doby, Tibor. *Discoverers of Blood Circulation: From Aristotle to the Times of Da Vinci and Harvey.* New York: Abelard-Schuman Ltd., 1963.

Hall, Marshall. *Researches Principally Relative to the Morbid and Curative Effects of Loss of Blood.* Philadelphia: E. L. Carey and A. Hart, Publishers, 1830.

Mathé, Jean. *Leonardo Da Vinci: Anatomical Drawings.* New York: Crown Publishers, 1978.

Morens, David M. "Death of a President." *New England Journal of Medicine* 341, no. 24 (December 9, 1999): 1845–1849.

Morgan, John. "Was Washington's Death Malpractice?" *USA Today on the Web,* February 22, 2000, and May 31, 2002. www.usatoday.com.

Siegel, Rudolph E. "Galen's Concept of Bloodletting in Relation to His Ideas on Pulmonary and Peripheral Blood Flow and Blood Formation." *Science, Medicine and Society in the Renaissance: Essays to Honor Walter Pagel.* Volume One. Edited by Allen G. Debus. New York: Science History Publications, 1972.

THREE

Amber, R. B., and A. M. Babey-Brooke. *The Pulse in Occident and Orient.* New York: Santa Barbara Press, 1966.

Broadbent, William Henry. *The Pulse.* Oceanside, N.Y.: Dabor Science Publications, 1977. Reprint of the 1890 edition published by Cassell & Company, Ltd., London.

Huang Ti Nei Ching Su Wen (The Yellow Emperor's Classic of Internal Medi-

cine). New Edition. Translated by Ilza Veith. Berkeley: University of California Press, 1966.

The Life of Sir William Broadbent. Edited by M. E. Broadbent. London: John Murray, 1909.

McCloud, Scott. *Understanding Comics: The Invisible Art.* Northampton, Mass.: Kitchen Sink Press, 1993.

Naqvi, N. H., and M. D. Blaufox. *Blood Pressure Measurement: An Illustrated History.* New York: Parthenon Publishing Group, 1998.

Nuland, Sherwin B. *The Mysteries Within: A Surgeon Reflects on Medical Myths.* New York: Simon & Schuster, 2000.

Seidel, Henry M., et al. *Mosby's Guide to Physical Examination.* Third Edition. St. Louis: Mosby, 1995.

Zimmerman, Leo M., and Katharine M. Howell. "History of Blood Transfusion." *Annals of Medical History* IV, no. 5 (September 1932): 415–433.

Four

Angier, Natalie. *Woman: An Intimate Geography.* New York: Houghton Mifflin Co., 1999.

Butler's Lives of the Saints. New Concise Edition. Edited by Michael J. Walsh. Great Britain: Burns & Oates Ltd., 1991.

Dean-Jones, Lesley. *Women's Bodies in Classical Greek Science.* Oxford: Oxford University Press, 1994.

Delaney, Janice, Mary Jane Lupton, and Emily Toth. *The Curse: A Cultural History of Menstruation.* Revised Edition. Urbana and Chicago: University of Illinois Press, 1988.

Frazer, James. *The Golden Bough.* Abridged Edition. New York: Penguin Books, 1996.

The Holy Bible, Revised Standard Version. New York: Meridian, 1974.

Museum of Menstruation Web site. February–April 2002. www.mum.org.

Pinkson, Thomas. "Sacred Feminine." Nierica—The Sacred Doorway Web site. March 30, 2002. www.nierica.com.

Raymond of Capua. *The Life of St. Catherine of Siena.* Translated by George Lamb. London: Harvill Press, 1960.

Ross, John Alan. "Plateau." *Handbook of North American Indians.* Volume 12. Edited by Deward E. Walker Jr. Washington, D.C.: Smithsonian Institution Press, 1998.

Teit, James A., and Franz Boas. *The Salishan Tribes of the Western Plateaus.* Extract from 45th B.A.E. Annual Report, 1927–1928.

Teresa of Avila. *The Life of Saint Teresa of Avila by Herself.* Translated by J. M. Cohen. New York: Penguin Books, 1957.

————. *The Life of Saint Teresa of Avila.* Translated and edited by E. Allison Peers. Catholic First Web site. February 28, 2002. www.catholicfirst.com.

"Women Were Considered Ritually Unclean." Women Priests Web site. April 10, 2002. www.womenpriests.org.

FIVE

"Antony van Leeuwenhoek." UC Berkeley Museum of Paleontology Web site. August 22, 2002. www.ucmp.berkeley.edu/history/leeuwenhoek.

Dobell, Clifford. *Antony van Leeuwenhoek and His "Little Animals."* New York: Harcourt, Brace and Co., 1932.

Jones, Thomas E. History of the Light Microscope Web site, © 1997. April 15, 2002. www.utmem.edu/~thjones/.

Schierbeek, Abraham. *Measuring the Invisible World.* London and New York: Abelard-Schuman, 1959.

Shinn, Al. Interview with author. Berkeley, Calif., August 28, 2002.

Turner, Gerard L'E. *Collecting Microscopes.* New York: Mayflower Books, 1981.

Verrier, Nancy Newton. *The Primal Wound: Understanding the Adopted Child.* Baltimore: Gateway Press, 1993.

SIX

Bäumler, Ernst. *Paul Ehrlich: Scientist for Life.* Translated by Grant Edwards. New York: Holmes & Meier, 1984.

Dr. Ehrlich's Magic Bullet, 1940, 103 minutes. Warner Bros. Pictures Video. Directed by William Dieterle.

Hirsch, James G., and Beate I. Hirsch. "Paul Ehrlich and the Discovery of the Eosinophil." *The Eosinophil in Health and Disease.* Edited by Adel A. F. Mahmoud, et al. New York: Grune & Stratton, 1980.

Marquardt, Martha. *Paul Ehrlich: Als Mensch und Arbeiter.* Berlin: Deutsche Verlags-Anstalt, Stuttgart, 1924.

———. *Paul Ehrlich.* New York: Henry Schuman, 1951.

Paul Ehrlich Institute Web site. January 9, 2002. www.pei.de.htm.

Pavao, Joyce Maguire. *The Family of Adoption.* Boston: Beacon Press, 1998.

Rhoads, Cornelius P. "Paul Ehrlich in Contemporary Science." *Bulletin of the New York Academy of Medicine* 30, no. 12 (December 1954): 976–987.

Silverstein, Arthur M. *Paul Ehrlich's Receptor Immunology: The Magnificent Obsession.* San Diego: Academic Press, 2002.

SEVEN

Bäumler, Ernst. *Paul Ehrlich: Scientist for Life.* Translated by Grant Edwards. New York: Holmes & Meier, 1984.

Doyle, Sir Arthur Conan. *Sherlock Holmes: The Complete Novels and Stories.* Volume One. New York: Bantam Books, 1986.

Ehrlich, Paul. "On Immunity with Special Reference to Cell Life." Croonian Lecture, Royal Society of London, March 22, 1900. Theoretical Immunology Web site. January 11, 2003. www.crystal.biochem.queensu.ca/forsdyke/theorimm.htm.

Hoffmann, Georg, and Brendt Birkner. *The Blood Handbook.* Point Roberts, Wash.: Hartley & Marks, Publishers, 1991.

Silverstein, Arthur M. *Paul Ehrlich's Receptor Immunology: The Magnificent Obsession.* San Diego: Academic Press, 2002.

Steranko, James. *The Steranko History of Comics.* Volume One. Reading, Penn.: Supergraphics, 1970.

Winger, Edward. Interviews with author. San Leandro, Calif., October 9, 2001, and February 7, 2003.

EIGHT

Abate, Tom. "Hidden Epidemic: Researchers, Policymakers Debate Tactics in Battle Against Hepatitis C." *San Francisco Chronicle* Web site. August 2, 2001. www.sfgate.com.

Armstrong, Walter. "The Untouchables." *Poz* (July 2001): 40.

California Health and Safety Code, Section 118340. California Department of Health Services Web site. March 18, 2002. www.dhs.cahwnet.gov.

Carroll, Chuck. "Jail Term Ordered for Reusing Needles." *Mercury News* Web site. August 16, 2002. www.bayarea.com.

Kennedy, Lisa. "The Miseducation of Nushawn Williams." *Poz* (August 2000).

Orcoff, Jerry. Interviews with author. San Jose, Calif., July 18, 2002, and February 17, 2003.

Rohde, David. "A Health Danger from a Needle. . . ." *New York Times* (August 6, 2001): 1.

Sanderson, Dale. Interview with author. San Jose, Calif., February 26, 2003.

Seyfer, Jessie. "Former Clinic Worker Facing 5-Year Sentence." *Mercury News* Web site. June 5, 2002. www.bayarea.com.

NINE

American Porphyria Foundation Web site. April 2003. www.porphyria foundation.com.

Bankard, Bob. "The Dracula Guide." Philly Burbs Web site. March 2003. www.phillyburbs.com/halloween2001/dracula.

Baring-Gould, Sabine. *The Book of Were-Wolves*. New York: Causeway Books, 1973.

Belford, Barbara. *Bram Stoker*. New York: Alfred A. Knopf, 1996.

Eckstein, Gustav. *The Body Has a Head*. New York: Harper & Row, 1970.

Florescu, Radu R., and Raymond T. McNally. *Dracula: Prince of Many Faces*. Boston: Back Bay Books, 1989.

Grossman, Mary Kay. Interview with author. Spokane, Wash., August 20, 2003.

Lane, Nick. "Born to the Purple: The Story of Porphyria." *Scientific American* Web site. December 16, 2002. www.sciam.com.

———. "New Light on Medicine." *Scientific American* (January 2003): 38–45.

Lassek, A. M. *Human Dissection: Its Drama and Struggle*. Springfield, Ill.: Charles C. Thomas Publisher, 1958.

MacMillan Illustrated Animal Encyclopedia. Edited by Dr. Philip Whitfield. New York: MacMillan Publishing Co., 1984.

McNally, Raymond T. *Dracula Was a Woman: In Search of the Blood Countess of Transylvania.* New York: McGraw-Hill Book Co., 1983.

Persaud, T.V.N. *A History of Anatomy: The Post-Vesalian Era.* Springfield, Ill.: Charles C. Thomas Publisher, 1997.

Porphyria: A Royal Malady. London: British Medical Association, 1968.

Shelley, Mary. *Frankenstein.* Author's introduction, 1831 edition. Oxford: Oxford University Press, 1969.

Stoker, Bram. *Dracula.* Introduction by Leonard Wolf. New York: Signet Classic, 1965 and 1992.

TEN

Aronson, Theo. *Grandmamma of Europe: The Crowned Descendants of Queen Victoria.* Indianapolis and New York: Bobbs-Merrill Co., 1973.

Ingram, G.I.C. "The History of Haemophilia." *Journal of Clinical Pathology* 29 (1976): 469–479.

Mannucci, Pier M., and Edward G. D. Tuddenham. "The Hemophilias—From Royal Genes to Gene Therapy." *New England Journal of Medicine* 344, no. 23 (June 7, 2001): 1773–1779.

National Hemophilia Foundation Web site. June 2003. www.hemophilia.org.

Neveu, Cindy. Interviews with author. Berkeley, Calif., June 23 and 26, 2003.

Pope-Hennessy, James. *Queen Mary.* New York: Alfred A. Knopf, 1960.

Potts, D. M., and W.T.W. Potts. *Queen Victoria's Gene.* Great Britain: Sutton Publishing Ltd., 1995.

Pullum, Christine. Telephone interview with author. June 10, 2003.

Resnik, Susan. *Blood Saga: Hemophilia, AIDS, and the Survival of a Community.* Berkeley: University of California Press, 1999.

Shemophilia Web site. June 2003. www.shemophilia.org.

Zeepvat, Charlotte. *Prince Leopold: The Untold Story of Queen Victoria's Youngest Son.* Great Britain: Sutton Publishing Ltd., 1998.

ELEVEN

Action Comics #403, "Attack of the Micro-Murderer" (August 1971). Cary Bates (writer) and Swan & Anderson (artists). New York: DC Comics.

"Blood: Frequently Asked Questions." U.S. Food and Drug Administration Web site. July 29, 2003. www.fda.gov/cber/faq/bldfaq.

Dayton, Andrew. Telephone interview with author. October 24, 2000.

Harveston, Richard. Interviews with author. Blood Centers of the Pacific, Irwin Center, San Francisco, Calif., June 12 and July 30, 2003.

Leno, Mark. Interviews with author. San Francisco, Calif., September 25, 2000, and September 14, 2001.

TWELVE

Angier, Natalie. *Woman: An Intimate Geography.* New York: Houghton Mifflin Co., 1999.

"The Complete Book of Men's Health—Priapism." Tiscali Web site. December 2003. www.tiscali.co.uk.

Cox, Paul. *Glossary of Mathematical Mistakes.* New York: Wiley, 1993.

Friedman, David M. *A Mind of Its Own: A Cultural History of the Penis.* New York: The Free Press, 2001.

Lemonick, Michael D. "The Chemistry of Desire." *Time* (January 19, 2004). *Time* magazine Web site. January 2004. www.time.com.

Park, Alice. "Sexual Healing." *Time* (January 19, 2004). *Time* magazine Web site. January 2004. www.time.com.

Rice, Anne. *Interview with the Vampire.* New York: Ballantine Books, 1976.

Silverton, Peter. "The Secret Life of Your Body." *The Observer* (November 25, 2001). *The Observer* Web site. January 21, 2004. www.observer.guardian.co.uk.

THIRTEEN

Barnett, Susan. Interview with author. Emeryville, Calif., September 3, 2003.

"Care for a Cure?" *Poz* (January 2001): 38–43.

Cimons, Marlene. "U.S. Officials Criticize Hyperthermia AIDS Treatment." *Los Angeles Times* (September 5, 1990): 13.

Homer. *The Odyssey.* Translated by E. V. Rieu. London: Penguin Books, 1991.

James, John S. "Hyperthermia Report: Only One Patient." *AIDS Treatment News* (June 1, 1990).

Levy, Jay. Interview with author. San Francisco, Calif., August 25, 2003.

Wilson, Keith D. *Cause of Death: A Writer's Guide to Death, Murder & Forensic Medicine.* Cincinnati: Writer's Digest Books, 1992.

ACKNOWLEDGMENTS

Many people helped bring this book to life. I would like to thank
the doctors and specialists who shared with me their expertise:
Dr. Susan Barnett, Rosemary Cozzo, Dr. Jay Gladstein, Mary Kay
Grossman, R.D., Richard Harveston, Dr. Shawn Hassler, Dr. Jay
Levy, Martin Pugh, and Dr. Edward Winger. Special thanks to
Dr. Donald Abrams for giving the final manuscript a thorough
medical checkup.

I am grateful to Cindy Neveu, Jerry Orcoff, Christine Pullum,
and Al Shinn for sharing their personal stories; to Steven Barclay
and Maurice Sendak for their support, loving as well as practical; to my fearless first draft readers, Jamie Inman and Lisa
Michaels; to Kathryn and Dan Mayeda for their faith and en-
couragement; and to Ballantine senior editor Dana Isaacson
and editorial assistant Deirdre Lanning for all their good work.
I consider myself fortunate for the continued guidance of
two wonderful women, my agent, Wendy Weil, and Ballantine
editor-in-chief Nancy Miller, whose thoughtful suggestions and
editing helped bring the manuscript to its final form.

I could never have started or completed this book, however,
were it not for Steve Byrne, my partner in life and in writing. He
gave his blood to my *Five Quarts,* literally and figuratively, and
made sure every last word here was polished and heartfelt and
true.

Page 10, statue of Perseus by Benvenuto Cellini, circa 1545: photo by the author; p. 24, diagram of Galen's conception of the circulatory system, from *The Discovery of the Circulation of the Blood* by Charles Singer, 1922: courtesy of the Kalmanovitz Library, University of California, San Francisco; p. 32, woodcut of woman applying leeches, 1639: courtesy of the National Library of Medicine; p. 50, portrait of William Harvey, from *The Discovery of the Circulation of the Blood* by Charles Singer, 1922: courtesy of the Kalmanovitz Library, University of California, San Francisco; p. 52, sixteenth-century engravings of the four temperaments by Virgil Solis: courtesy of the National Library of Medicine; p. 63, snapshot of the Hayes siblings: photo provided by the author; p. 88, portrait of Antoni van Leeuwenhoek: courtesy of the San Francisco Public Library (Picture File); p. 94, replica of a Leeuwenhoek microscope: courtesy of Al Shinn; p. 119, Be Here for the Cure sticker: courtesy of the San Francisco AIDS Foundation © 1992; p. 129, publicity shot from *Dr. Ehrlich's Magic Bullet,* 1940: from author's personal collection; p. 135, illustration of Holmes and Watson, from *The Strand,* vol. VI, July–Dec., 1893: courtesy of the San Francisco Public Library; p. 139, photo of Paul Ehrlich in his laboratory: courtesy of the Paul Ehrlich Institute; p. 164, butterfly needle: photo by the author; p. 168, *The Reward of Cruelty* by William Hogarth, 1751, from *The Works of William Hogarth,* vol. II, 1833: courtesy of the San Francisco Public Library; p. 174, engraving of blood transfusion operation, from *Scientific American,* Sept. 5, 1874: courtesy of the San Francisco Public Library; p. 180, portrait of Elizabeth Bathory: courtesy of Dennis Báthory-Kitsz (www.bathory.org); p. 194, photo of Queen Victoria and Prince Leopold, 1862: The Royal Archives © 2004 Her Majesty Queen Elizabeth II; p. 228, "Action Comics" #403 © 1971 DC Comics. All Rights Reserved. Used with Permission.; p. 233, "Your Blood Can Save Him" poster: courtesy of the American Red Cross Museum. All rights reserved in all countries.; p. 241, engraving of Leonardo da Vinci: courtesy of the San Francisco Public Library (Picture File); p. 257, photo of Jay Levy, 1984: courtesy of Dr. Jay Levy; p. 263, illustration of Odysseus by Jacomb Hood, from *The Boy's Odyssey* by Walter C. Perry, 1901: courtesy of the San Francisco Public Library.

INDEX

Action Comics, 227–28
Addison, William, 126–27
adoption, adoptions, 106–7, 109–13
   family trees and, 115
   seeking blood kin and, 110–13
Aeneid, The (Virgil), 6
African Americans, 233–34
Against Erasistratus (Galen), 18–19
AIDS, *see* HIV/AIDS
American Red Cross, 232–34
American Sign Language, 41
Anatomical Essay on the Movement of the Heart and Blood in Animals, An (Harvey), 49
Anecdotes of the Habits and Instincts of Animals, 182
anemia, 33, 127
   porphyria and, 183, 187–89
   sickle-cell, 103, 243
   transfusions and, 51
Angier, Natalie, 244
Animal Spirits, 25
antibodies, 125, 245
   blood types and, 221–22
   Ehrlich and, 138–41, 152
   HIV/AIDS and, 229, 231, 254
   lymphocytes and, 132, 141
antigens, 147, 221–22
antitoxins, 138–41
apheresis, 224
Apollo, 5
Aristotle, 15–16, 67
arteries, 7, 15–16, 20, 25, 47–51, 88, 212
   blood pressure and, 45
   carotid, 41
   in eyes, 236
   Harvey on, 49–50
   Leeuwenhoek and, 97
   pulmonary, 16
   pulse and, 41–45, 47, 50
   radial, 42, 173
   sex and, 240, 243
   transfusions and, 51, 173
Asclepius, 5–6, 10, 23, 258
Athena, 5, 10

Bacon, Roger, 90
bacteria, 127, 133, 137, 206, 246
   Leeuwenhoek and, 87, 89, 96–97
   lymphocytes and, 131
barbers, barbershops, bloodletting by, 28, 168
Barnett, Susan, 258–62
basophils, 127
Bathory, Elizabeth, 179–82, 184
Bay Area Reporter, 117
B cells, 132, 141
   HIV/AIDS treatments and, 254, 259–60
Be Here for the Cure campaign, 118–19, 253, 257
Behring, Emil von, 139
birth control pills, 201, 208
black bile, 16, 25, 52
blood:
   in beauty regimens, 180–81
   cash value of, 4, 219
   clotting of, 100, 172–73, 182, 193, 195–96, 200–202, 205–6, 210–11, 216
   consumption of, 4–5, 181, 187–88
   distances traveled by, 7, 213, 236
   donations of, 214–15, 217–18, 223–35
   fear of, 33
   in mythology, 5–6, 9–11, 261–64
   odor of, 4
   rare, 222
   speeds traveled by, 212
   storage of, 213–16, 219, 222–23
   taste of, 3–4, 7

blood (*cont'd*):
testing of, 14, 26–31, 33–37,
127, 143–56, 159–60,
163–64, 166, 215–16, 218,
225, 229–30, 250
transfusions of, 6, 51–54, 86–87,
171–76, 183, 187, 200, 217,
219–20, 222–24, 227–28,
232
types of, 172, 220–22
*Blood: An Epic History of Medicine
and Commerce* (Starr), 233–34
blood banks, 4, 213–26
and ban on gay donors, 229–35
blood types and, 220–22
donor screening and, 218,
225–26
errors by, 234
finances of, 219, 224
history of, 233–34
storage in, 215–16, 219, 222–23
testing in, 215–16, 218, 225
blood-brain barrier, 252–53
Blood Centers of the Pacific,
213–26, 231–32
blood kin, 110–15
adoption and, 106–7, 110–13
family trees and, 114–15
bloodletting, 26–29, 173
by barbers, 28, 168
Galen and, 14, 17–21, 23, 29, 48
leeches in, 31–33
in Middle Ages, 27–28
porphyria and, 188–89
precedents in nature for, 21
tools of, 31–32
blood pressure, 45–48, 58
arteries and, 45
Broadbent on, 43, 45–46
high, 45, 108–9
low, 45–46
sex and, 244–45
blue blood, 191
Body Has a Head, The (Eckstein),
169–70
Body in Question, The (Miller),
50–51
bone marrow, 103, 132
brain, 25, 244, 252–53, 265

breasts, 240
Brehmer, Hermann, 133
Broadbent, Sir William Henry:
on blood pressure, 43, 45–46
pulse and, 42–47
Byrne, Steve, 8–9, 24, 48, 114–16,
118–19, 235, 246–51, 264
blood tests of, 14, 26–31, 34–37,
143–46, 148–54, 156, 250
comic books and, 13, 54–56,
84–85, 148, 227
family tree of, 114
Giorgi case and, 36–37, 153–56,
159, 161–62, 165–66
medications and, 13–14, 37–39,
56, 246–47, 250
Byron, Lord, 177–78

cadaver dissections, 167–69
California, University of, at San
Francisco, 251–53
cancer, 120, 232
capillaries, 7, 50, 103, 237
Leeuwenhoek and, 88, 97
sex and, 240, 247
Captain America, 86
carbon dioxide, 103
cardiovascular disease, 47
Carmelites, 81–83, 104, 109
Carmilla (Le Fanu), 177
carotid artery, 41
Catherine of Siena, Saint, 77, 79–80
CD8 Antiviral Factor (CAF),
255–56, 261
Centers for Disease Control, 235
Charité Hospital, 125, 128, 130
Charlotte, Princess, 189
chemokines, 261
chemotherapy, 120, 128
children, venesection in, 23
China, pulse taking in, 44–45
Chiron, 5, 258
Chiron Corporation, 258–61
Christ, 76–77
blood of, 76
crucifixion wound of, 79–80
Christianity, 26, 179, 191
menstruation and, 70
Chrysaor, 11

Circe, 263–64
circulatory system, 190, 212–13,
    223
  blood clots and, 172
  blood kin and, 113
  Galen and, 15–16, 24, 49–51
  Harvey on, 49–51
  porphyria and, 187
  sex and, 240, 243
  transfusions and, 52
circumcision, 196
Circus Maximus, 18
clitoris, 240, 244
clotting, *see* blood, clotting of
cocaine, Ehrlich's research on,
    137–38
comic books:
  blood drives in, 227–28
  origin stories in, 85–87, 104
Commodus, Emperor of Rome, 22
complete blood counts (CBCs), 127
compound microscopes, 91, 102,
    143
Conant, Marcus, 253
Corpus Iuris Canonici, 78
Cozzo, Rosemary, 29–31, 33–34
crying, 26
cryoprecipitate, 202–6, 209–11,
    214
cupping, 31, 33, 188
curse of Eve, 70
Cyclopes, 262–63
cytochemistry, 125
cytokines, 132
cytomegalovirus (CMV), 217

Dayton, Andrew, 232, 234–35
death, 265
  absence of pulse in, 41
  in comic books, 56
  from HIV/AIDS, 119
defibrination, transfusions and,
    172–73
degli Armati, Salvino, 91
De Insomniis (Aristotle), 67
Denis, Jean-Baptiste, 52–53
Dialogue of Saint Catherine, The
    (Catherine of Siena), 79–80
diphtheria, 137, 139–40, 198

dissections, 167–69
DNA, 104, 191, 220
  blood kin and, 113
  blood tests and, 147–50
  HIV/AIDS treatments and, 254,
    259
Dr. Ehrlich's Magic Bullet, 121–22,
    129
Dolphin, David, 184, 187
Doyle, Sir Arthur Conan, 134–37
Dracula (Stoker), 170–80, 182–83,
    237–38
  realism in, 183
  precedents for, 176–80, 182–83
  transfusions in, 171–76, 183
Dracula Was a Woman (McNally),
    179–80

e-chairs, 223–25, 249
Eckstein, Gustav, 169–70
Edward, Duke of Kent, 193
Egypt:
  bloodletting in, 17
  Ehrlich and, 130, 134
  pulse taking in, 44
Ehrlich, Hedwig Pinkus, 128–30,
    133–34
Ehrlich, Marianne, 128–29
Ehrlich, Paul, 120–30, 132–42
  colors and, 123–24, 130
  hematology and, 126–28
  immunity and, 138–42
  Marquardt on, 122–23, 134,
    142
  reading preferences of, 134–36
  Royal Society and, 140–41
  stains used by, 124–27, 129–30,
    136–37, 152
  syphilis cure and, 120–22, 125
  TB of, 130, 133, 135–36
  toxin research of, 137–41
Ehrlich, Stefanie, 128
electron microscopes, 142–43
ELISA-HIV test, 150, 229, 231
Elsholtz, Johann, 53
emotions, 7
  pulse and, 43, 48
endorphins, 12–13, 117, 245
eosinophils, 127

Erasistratus, 18–21, 48
erections, 240–44
erythroblasts, 103–4
erythrocytes, *see* red blood cells
exsanguination:
  measurements in, 169–70
  see also vampires, vampirism
eyes, 236–37

family trees, 114–15
Fantastic Four, 54, 84–85
fevers, 246, 249
fibrin, fibrinogen, 172–73
  deficiency in, 204–6, 208–10
Flash, 13
Flow Cytometer, 151–52
Food and Drug Administration
  (FDA), 4, 206, 229–32
  blood donor policies and,
    229–30, 232, 234–35
  HIV/AIDS treatments and,
    249–50, 258
Frankenstein (Shelley), 178
Frazer, James, 65–68
Frederick I, King of Prussia, 97
Frerichs, Friedrich von, 125, 129
Friedman, David M., 242
Fuseli, Henry, 70

Galen, 14–26, 33, 48–51
  bloodletting and, 14, 17–21, 23,
    29, 48
  circulatory system and, 16,
    24–25, 49–51
  cosmetics and, 22–23
  factual errors of, 24–25, 49–51
  humors and, 16–17, 19, 21
  self-confidence of, 23–24
  on spirits, 24–26, 50
gallbladder, 17, 24
garlic, 186–87
gays, *see* homosexuals,
  homosexuality
gender, gender differences:
  blood transfusions and, 172
  Catholic Church and, 78–79
  hemophilia and, 193, 195–96,
    199–203, 208
  sex and, 240–45

  see also homosexuals,
    homosexuality
Genesis, 70
genetics, 113–15
  family trees and, 114–15
  hemophilia and, 193, 196,
    198–99, 202, 205
  porphyria and, 185–86
  see also DNA
George, Prince of Wales, 46
George III, King of England,
  185–86, 188–89
George IV, King of England, 189
Gerhardt, Carl, 129–30
Giorgi, Elaine, 154–66
  bizarre behavior of, 159–60,
    162
  sentencing of, 161–66
  syringes reused by, 154, 157,
    159–60, 164
gladiators, 187
  blood drinking of, 4–5
  Galen and, 14–17
Godwin, Claire, 177
Golden Bough, The (Frazer), 65–67
Gordon, Ruth, 129
Gorgon, 5–6, 9–11
Graaf, Reinier de, 95–96, 242–43
Grossman, Mary Kay, 187–88
growth hormone, 132
Guild of Barbers and Surgeons,
  168

Hall, Marshall, 33
Hart, Mary, 85
Harveston, Richard, 213–25
Harvey, William:
  on circulatory system, 49–51
  transfusions and, 53
Hassler, Shawn, 144
Hayes, Colleen, 60–61, 63, 72
Hayes, Ellen, 60–64, 77, 80
Hayes, Julia, 60–63
Hayes, Maggie, 60–64, 72, 74, 81
Hayes, Shannon, 60–65, 69–83,
  104–11
  adoption and, 106–11
  author's childhood and, 60–65,
    69–70, 72–77, 111

gynecological problems of, 70–72, 74
pregnancy of, 105–9
spirituality of, 75–83, 104–5, 109
heart, 6, 15–16, 34, 80, 116, 212, 243–45, 265
blood kin and, 113
Galen on, 50–51
Harvey on, 49–51
Leonardo on, 25–26
pulse and, 41, 43–45, 47–48
sex and, 244–45
transfusions and, 173–74
Helena, Princess of Waldeck, 197
hematology, 137, 152
blood banks and, 220
Ehrlich and, 126–28
hemophilia and, 195
hemochromatosis, 189
hemoglobin, 102–4, 127, 135–36, 184, 218
hemophilia, 112, 172, 189, 192–203, 205
fibrinogen deficiency and, 205
naming of, 199
prevalence of, 195–96
support groups for, 208–10
treatments for, 196–97, 200–203, 205, 210, 252
hemophobia, 33
hemorrhoids, 21
Henry, Prince of Prussia, 198
Henry VIII, King of England, 168
hepatitis, 36, 158, 203, 215
hepatitis C virus (HCV), 206
hemophilia and, 203
syringe reuse and, 157–59
Hermes, 10
Herophilus, 48
Hewson, William, 126
Hippocrates, 15–16, 18, 23, 48
Hippolytus, 6
Hirudo medicinalis, 32
Hitchcock, Alfred, 176
HIV/AIDS, 8–9, 115, 117–20, 238, 246–62
Barnett and, 258–62
bleeding disorders and, 203, 206, 209

blood donations and, 215, 217–18, 229–35
criminal acts and, 36, 153–55, 159–60, 165
deaths from, 108, 117–19
educational campaigns and, 118–19, 253, 257
immune system and, 143, 151, 250, 253–55, 260
Levy and, 251–59, 261–62
medications and, 13–14, 27, 37, 39, 56, 119, 143, 148, 151, 155, 210, 235, 246–54, 256
microscopy and, 143, 151
Muscle System and, 117–18
support groups for, 209
syringe reuse and, 36, 39–41, 153–54, 164–65
testing and, 14, 26–31, 34, 36, 143–44, 147–52, 154, 229–31, 233
vaccine, 258–60
HIV and the Pathogenesis of AIDS (Levy), 252
HMOs, 48
Hogarth, William, 168–69
Holmes, Sherlock, 135–37
Homer, 262–64
homosexuals, homosexuality, 55, 116, 242
of author, 81, 83, 104–5
blood donations and, 228–35
family trees and, 114
HIV/AIDS and, 34, 119, 209, 253
Hooke, Robert, 91
Huang Ti Nei Ching Su Wen (The Yellow Emperor's Classic of Internal Medicine), 44–45
Hulk, 39, 86–87
Human Torch, 86
humors, 16–17, 24–25, 52–53, 188
bloodletting and, 17, 19, 21, 28
Galen and, 16–17, 19, 21
transfusions and, 52–53
Huston, John, 121
hypertension, 45, 108–9
hyperthermia, 249
hysterectomies, 71–72, 74

ichor, 264
immunity, immune system, 217, 246
    Ehrlich and, 138–42
    HIV/AIDS and, 143, 151, 250,
        253–55, 260
    side-chain theory of, 140–42
Immunodiagnostic Laboratories
    (IDL), 28–29, 35
    author's tour of, 144–52
immunoglobulin, 245
Insulin-Resistance Diet, The
    (Grossman), 187
Interview with the Vampire (Rice),
    238–39
Inuits, 188
Invisible Woman, 85, 148
iron, 4, 103, 188–89
Irving, Henry, 176
Iscador, 250

Jansen, Zacharias, 91
jaundice, 158
Jenner, Edward, 138
Jews, Judaism, 124
    circumcision of, 196
    menstruation and, 69
Johnson, Magic, 148

Karimojong, 188
kidney disease, 45
kiveris vein, 25
Kolosh Indians, 65–66
Krim, Mathilde, 253

"L'Amour" (Michelet), 67
Landsteiner, Karl, 220–21
Lane, Nick, 186
Larousse Gastronomique, 188
laxatives, 19–20
leeches, 31–33, 188
Leeuwenhoek, Antoni van, 87–102,
    126
    errors of, 100–101
    humble beginnings of, 88–89
    measurements of, 100
    microscope of, 87, 89–102
    Royal Society and, 95–97,
        99–101, 140
    wild experiments of, 96

Leeuwenhoek, Maria van, 89,
    92–93
Le Fanu, J. Sheridan, 177
Leno, Mark, 231–32
Leonardo da Vinci, 25–26,
    241–42
Leopold, Prince, Duke of Albany,
    192–98
leukemia, 127, 224
leukocytes, *see* white blood cells
Leviticus, 69, 78
Levy, Jay, 251–59, 261–62
    HIV discovered by, 251–52
    HIV treatments and, 252–56,
        258–59, 261–62
liver, 16, 25, 49
Lives of the Saints, The, 77
Lower, Richard, 51
love sickness, diagnosis of, 48
love vein, 26
lungs, 16, 102–3, 116
lymphocytes, 127, 131–32, 255
    antibodies and, 132, 141
    life span of, 213
    physical appearance of, 152

McNally, Raymond, 179–80
mad cow disease, 225
magic bullets, 119–22, 125
Marcus Aurelius, Emperor of Rome,
    21–22
Marquardt, Martha, 122–23, 134,
    142
Marti, Inmaculada, 250
Mary, Queen of Scotland, 186
Mary II, Queen of England, 97
Masai, 188
mast cells, 124
Matthews, Brian, 154, 156, 162,
    164, 166
Maupin, Armistead, 116, 238
medications, 33, 51, 202–7,
    209–11, 214, 217
    blood clots and, 172–73, 182
    Byrne and, 13–14, 37–39, 56,
        246–47, 250
    Ehrlich and, 120–22, 125, 128
    for fibrinogen deficiency, 204–6,
        211

Galen and, 19–20, 22
hemophilia and, 202–3, 205, 210, 252
HIV/AIDS and, 13–14, 27, 37, 39, 56, 119, 143, 148, 151, 155, 210, 235, 246–54, 256
magic bullets and, 119–22, 125
naming of, 250–51
sex and, 243–44, 246–47
for transverse myelitis (TM), 207
Medusa, 5, 10–11
memory cells, 259–60
menopause, 162
menstruation, 7, 63–74
author's childhood and, 63–65, 69–70, 72–74, 111
birth control pills and, 201, 208
bloodletting inspired by, 21
hemophilia and, 200–201
irregular, 82, 105
Leonardo on, 25
religions and, 69–70, 78–79, 82
seclusion during, 65–67
in social contexts, 68–69
superstitions about, 67–68
Michelet, Athenais, 67
Michelet, Jules, 67
Micrographia (Hooke), 91
microscopes, microscopy, 141–43
compound, 91, 102, 143
Ehrlich and, 124–30, 141
electron, 142–43
in history, 90–91
HIV/AIDS and, 143, 151
of Leeuwenhoek, 87, 89–102
Middle Ages, 23, 67–68, 181, 186
blood kin and, 113
bloodletting in, 27–28
milk vein, 25
Miller, Jonathan, 50–51
Mind of Its Own, A (Friedman), 242
Minigawa, Rahn, 162–63
Molyneux, Thomas, 90
Mullin, Hugh F., III, 161–62, 164–65
Muscle System, 115–19
HIV/AIDS and, 117–18
Muses, 11, 262

Nabokov, Vera, 8
Nabokov, Vladimir, 7–8
National Institutes of Health, 235
Native Americans, 65–66, 68–69, 73, 188
Natural History (Pliny), 67–68
Natural Spirits, 25, 241
Nero, Emperor of Rome, 90
neutrophils, 127
Neveu, Cindy, 203–11, 214
Newsweek, 193
Newton, Sir Isaac, 95, 98, 140
Nicholas II, Czar of Russia, 198–99
Nietzsche, Friedrich, 134
Nightmare, The (Fuseli), 70–71
nosebleeds, 21
Nymphs, 10

Odyssey, The (Homer), 262–64
"On Immunity with Special Reference to Cell Life" (Ehrlich), 140–41
Opus (Bacon), 90
Orcoff, Jerry, 157–58, 161–62, 165–66
orgasms, 244–45
oxygen, 108, 116, 184
sex and, 240, 243
transportation and discharge of, 102–4
oxytocin, 244–45

Panacea, 6, 258
Pavao, Joyce Maguire, 115
Pegasus, 11
penis, 25, 240–44
Pergamum, 15, 18
Perseus, 10–11
Peter I, the Great, Czar of Russia, 97–98
phlebotomy, *see* bloodletting
phlegm, 16, 52
placenta, 108
plant cells, 91
plasma, 4, 139, 190
blood banks and, 214, 216, 218–19, 224
fibrinogen deficiency and, 205
hemophilia and, 200, 202, 210

plasma (*cont'd*):
  Leeuwenhoek and, 100
platelets, 4, 171–72, 190, 195,
    213–16
  blood banks and, 214–16,
    218–19, 224
  life span of, 213, 216
Plato, 25
Pliny the Elder, 67–68, 90
Polidori, John, 177–78
porphyria, 183–89
Porter, Chris, 57–60
Praxagoras, 48
preeclampsia, 108
pregnancy, 25
  of author's sister, 105–9
  hemophilia and, 202–3
priapism, 243
Priapus, 243
Primal Wound, The (Verrier), 110
Project Inform, 249–50
protease inhibitors, 250–51
protozoans, 87, 89, 96
Psycho, 176
p24 antigen test, 147
Pullum, Christine, 199–203, 209
Pullum, Doyle, 199, 201–3
pulmonary artery, 16
pulse:
  in ancient history, 44–45, 48
  of author, 46–47
  Broadbent and, 42–47
  as diagnostic tool, 43–45, 47–48
  Harvey on, 50
  taking of, 41–48
Pulse, The (Broadbent), 42–44, 46
purgatives, 19–20

Q-PCR test, 147–49

radial artery, 42, 173
red blood cells, 4, 33, 100–104,
    126–27, 190–91
  of author, 101–2
  blood banks and, 216, 218–24
  blood types and, 220–21
  Leeuwenhoek and, 87–88,
    94–95, 100–101, 126
  life span of, 213, 216

nuclei absent from, 104, 191
  origin of, 103–4
  oxygen transported and
    discharged by, 102–4
  physical appearance of, 100, 103
  transfusions and, 51
religion, 6–7, 55, 75–80, 171
  Galen and, 23
  menstruation and, 69–70, 78–79,
    82
  vampirism and, 175–76, 237
  see also Jews, Judaism
Renaissance, 32, 49, 180, 241
rest cure, 133–34
Reward of Cruelty, The (Hogarth),
    168–69
Rh factor, 221–22, 228
Rice, Anne, 237–39
ricin, Ehrlich's research on, 138,
    140
RNA, 254–55
Robinson, Edward G., 121
Roman Empire, 18, 21–22
Royal Institute of Experimental
    Therapy, 139–40
Royal Society of London:
  Ehrlich and, 140–41
  Leeuwenhoek and, 95–97,
    99–101, 140
Rymer, James Malcolm, 177

St. Mary's Hospital, 43, 45
saliva, Leeuwenhoek on, 96
Salvarsan, 120–21
Sanderson, Dale, 159–64
San Francisco AIDS Foundation,
    119, 253
Savage Dragon, 87
scavenger cells, 103
Schönlein, Johann, 199
Sekhmet, 44
selective staining, 124
semen, 9, 87, 113, 243
Seneca, 90
sex, sexuality, 9, 240–47
  blood donations and, 228–31,
    235
  clitoris and, 240, 244
  erections and, 240–44

HIV/AIDS and, 155, 158,
    246–47
  illnesses and, 245–47
  orgasms and, 244–45
  of vampires, 176, 240
  see also gender, gender differences;
    syphilis
Shakespeare, William, 49, 176, 197
She-Hulk, 86–87
Shelley, Mary Godwin, 177–78
Shelley, Percy, 177–78
Shemophilia.org, 208–10
Shinn, Al, 101–2
  Leeuwenhoek's microscope and,
    92–95, 98–99, 101
Shoshoni Indians, 68–69
sickle-cell anemia, 103, 243
side-chain theory of immunity,
    140–42
Sign of Four, The (Doyle), 137
Simpson, O. J., 154
sleep, 131–34
  lymphocytes and, 131–32
  TB and, 133–34
smallpox, 138
SmithKline Beecham, 35–36,
    153–54, 157, 159, 163–64
Solis, Virgil, 52
soul, link between blood and, 6–7,
    265
spectacles, invention of, 90–91
sperm, sperm cells, 25, 87, 89
sphygmographs, 43–44
Spider-Man, 84
Spider-Woman, 86
spirits, 6, 88, 241
  Galen on, 24–26, 50
Spitfire, 86
spleen, 25, 103
Spokane Indians, menstruation and,
    68, 73
spring lancets, 31
stains, staining:
  Ehrlich and, 124–27, 129–30,
    136–37, 152
  selective, 124
  vital, 129, 136–37
Starr, Douglas, 233–34
starvation, 19

stem cells, 103
Stocking, George W., Jr., 66
Stoker, Bram, *Dracula* and, 170–80,
    182–83, 237–38
Stoker, Thornley, 183
Study in Scarlet, A (Doyle), 135–36
Superman, 55, 85–86, 227–28
support groups, 203, 208–10
syncope, bleeding to, 33
synesthesia, 7–8
syphilis:
  blood banks and, 215
  medications for, 120–22, 125
syringes:
  blood testing and, 33–34, 36–37,
    153–54, 163–64
  in history, 31, 51
  reuse of, 36–41, 150, 153–54,
    157–60, 164–65
  transfusions and, 51
  for vitamin B_{12} injections, 37–38

Tales of the City (Maupin), 116, 238
T cells, 131–32
  blood tests and, 150–52
  HIV/AIDS and, 143–44, 247,
    249–50, 254–55, 259–60
Teck, Prince, 192
Teiresias, 263
temperaments, 52–53
Teresa of Avila, Saint, 77–79, 81
testosterone, 117
Thérèse of Lisieux, Saint, 77
Tibetans, 188
Time, 143
toxins, 184
  Ehrlich's research on, 137–41
  hepatitis and, 158
transubstantiation, miracle of,
    76–77
transverse myelitis (TM), 206–7
tuberculosis (TB), 133–34, 136–37
  of Ehrlich, 130, 133, 136
  rest cure for, 133–34
Tuberculosis Is a Curable Disease
    (Brehmer), 133
typhoid fever, 46

Utrecht Museum, 92

vampire bats, 182
vampires, vampirism, 170–89,
    237–40
  of Bathory, 179–82, 184
  blood drive of, 240
  in literature, 170–80, 182–83,
    237–39
  medical basis for, 183–89
"Vampyre, The" (Polidori), 177–78
Varney the Vampire (Rymer), 177
Vega, Suzanne, 167
veins, 5–7, 15–21, 37, 48, 84, 86,
    88, 212
  in bloodletting, 17–21, 23,
    27–33
  for blood test draws, 29–30
  eyes and, 236–37
  Harvey on, 50
  kiveris, 25
  Leeuwenhoek and, 97
  Leonardo on, 25–26
  love, 26
  milk, 25
  sex and, 243–44
  transfusions and, 172–75
vena amoris, 26
Vermeer, Jan, 87, 89
Verrier, Nancy Newton, 110
Vesalius, Andreas, 49
Victoria, Queen of Great Britain, 42,
    46, 189, 191–94, 197–98
viral load testing, 146–51
Virgil, 6
virgins, virginity, 7, 180–81
viruses, 46, 127, 138, 217, 246
  lymphocytes and, 131–32

  see also hepatitis; hepatitis C virus;
    HIV/AIDS
Vital Spirit, Vital Spirits, 6, 25–26,
    50
vital staining, 129, 136–37
vitamin B_{12} injections, 37–38
von Willebrand's disease, 195, 208

Waldeyer, Wilhelm, 124
Washington, George, 33
Weigert, Carl, 124
white blood cells, 4, 131–32, 190,
    246
  of author, 102
  blood banks and, 216–18
  Ehrlich and, 126–27
  functions of, 126–27
  HIV/AIDS and, 144, 254–55,
    260
  Leeuwenhoek and, 100
  life span of, 213
  physical appearance of, 152
white coat syndrome, 58
Williams, Nushawn, 155
Winger, Edward, 145–52
Wintrobe, Maxwell, 137
Woman: An Intimate Geography
    (Angier), 244
World War II, 121, 233–34
Wren, Christopher, 51

X-Men, 54–56, 225

yellow bile, 16, 52–53

Zeus, 6

FIVE QUARTS

BILL HAYES

A Reader's Guide

1. In *Five Quarts,* Hayes traces the changing attitudes toward blood over time, from classical antiquity to the age of AIDS. Discuss how perceptions of blood have changed within your lifetime, both for yourself personally and for the culture at large. Also, how did your perceptions change over the course of reading this book?

2. Squeamishness is highly subjective, and for some, simply reading about blood makes them want to reach for a Dramamine. Were there passages in *Five Quarts* that made you feel squeamish?

3. The author interweaves science and history with a memoir that is, at times, intensely personal. Did this style enhance or detract from your experience of reading the book? Did you find yourself preferring one narrative thread over another?

4. Hayes writes that blood was once thought to house the human soul, "a reverential perspective almost beyond imagination today, when blood is widely considered hazardous waste material." This is a strong position. How does the author's opinion mesh with yours?

5. While profiling figures ranging from England's Queen Victoria to scientists Paul Ehrlich and Antoni Van Leeuwenhoek to contemporary women living with hemophilia, Hayes brings out the human stories found in blood. Discuss a person (yourself, perhaps?) whose life has been affected or defined in some way by blood.

6. One of the more intimate stories told in the book concerns the author's sister Shannon, who as a young woman faced some daunting decisions: whether to enter cloistered life, whether to give up her child. If you were Shannon's sibling, how would you have counseled her?

7. *Five Quarts* includes a frank look inside the relationship of Hayes and his HIV-positive partner. How did you respond to this storyline?

8. Typically, a nonfiction author is the reader's trusty guide. What was your response when your guide admitted to being "lost," such as during Hayes's conversation with the director of the IDL blood lab in chapter 7 and with the AIDS vaccinologist in chapter 13?

9. In writing about contemporary blood banking, Hayes advocates changing the policy restricting healthy, HIV-negative gay men from donating blood. How did you respond to Hayes's argument? Do you agree or disagree with the policy?

10. Go to the Red Cross website (www.redcross.org/donate/give) and print out the donor eligibility guidelines. While reading through them, discuss the most surprising reasons why a person may be deferred, temporarily or permanently, from giving blood.

11. The blood-bathing countess Elizabeth Bathory went to horrific extremes in her quest to preserve her youth and beauty. Did her story make you think about the lengths people go to today to look youthful? What's the most outrageous thing you've done in the name of vanity?

12. Hayes's book is populated by a large cast of historical figures whose impact continues to be felt today, from the ancient Greek physician Galen to the seventeenth-century English anatomist William Harvey to *Dracula* author Bram Stoker. Which historical figures did you find especially intriguing?

13. The "renegade phlebotomist" from chapter 8 spent one year in a county jail. Did she receive an appropriate sentence for her actions?

14. Sharing a common belief of his day that qualities ran in the blood, seventeenth-century German surgeon Johann Elsholtz proposed a novel remedy for marital discord: a mutual blood transfusion between husband and wife. Hypothetically, which of your significant other's qualities would you least like to get in such a trade? Which would you most enjoy?

15. The author explains and dispels several ancient beliefs about blood. Are there myths or old wives' tales about blood that are alive and well in your family?

16. A recurrent theme of the book is "the search for a cure," and yet, history has shown, cures are very elusive. Are there diseases you believe will be cured in your lifetime?

17. Rate yourself according to the ancient doctrine of temperaments described in chapter 3. Do you tend to be a sanguine person (cheerful), melancholic (gloomy), choleric (irritable), or phlegmatic (lethargic)? How does your private assessment jibe with the group's perception of you?

PHOTO: © STEVE BYRNE

BILL HAYES is the author of the national bestseller *Sleep Demons: An Insomniac's Memoir*. His work has been published in *The New York Times Magazine* and *Details,* among other publications, and at Salon.com. He has also been featured on many NPR programs as well as the Discovery Health channel. He lives in San Francisco with his partner, Steve.